Paved Track Stock Car Technology

By Steve Smith

Editorial Assistant: Lorianne Twill

ISBN 0-936834-37-4

Revised April 2005

Published By

STEVE SMITH AUTOSPORTS® PUBLICATIONS

P.O. Box 11631 / Santa Ana, CA 92711 / (714) 639-7681
www.SteveSmithAutosports.com

Printed and bound in the United States of America

Table of Contents

Introduction

Paved track stock car racing has to be one of the most fun, exciting and challenging sports in America.

This book was written to help the competitors in this sport — no matter what class of competition he is involved in — to better understand the intricacies of race car preparation, chassis setup and adjustment, and to help keep the challenge exciting.

Technology and change create new challenges for racers — how to work with and understand them, how to go faster. This book, like all the others I have written, is meant to help racers meet those challenges. Our goal is to make life easier for the racer...to help him understand his race car so that racing can be an enjoyable, instead of a frustrating, experience. We want to help the racer put it all together for the optimum level of performance. Hopefully this book will fulfill that goal for you.

Have fun, be safe, and happy racing!

Steve Smith

A Very Special Thanks

No book of this size and magnitude could ever be written without the help and cooperation of many special people. I would like to extend a very special thanks to: George Gillespie and Gene Roberts of PRO Shocks, Gary Sigman of Professional Racers Emporium (PRE), Ken Fasola of Wilwood Engineering, Kenny Sapper of Speedway Engineering, Scott Keyser of AFCO, Tex Powell of Tex Racing Enterprises, Chas Howe of Howe Racing Enterprises, and Lizard Fonnegra of Dan Press Industries. Thank you! I sincerely appreciate all the time you took to help me.

Also, thank you to Gary Smith of Performance Design, a very talented artist, for another great book cover.

Most importantly, a very special thank you goes to my very understanding and dedicated family who have to make special allowances for me while I am researching and writing a book. This book is dedicated to Georgiann and Lori — the two most important people in my life. Thanks for your caring and loving support!

Steve Smith

Disclaimer Notice

Every attempt has been made to present the information contained in this book in a true, accurate and complete form. The information was prepared with the best information that could be obtained. However, auto racing is a dangerous undertaking and no responsibility can be taken by any persons associated with this book, the author, the publisher, the parent corporation, or any person or persons associated therewith, for injury sustained as a result of or in spite of following the suggestions or procedures offered herein. All recommendations are made without any guarantee on the part of the author or the publisher, and any information utilized by the reader is done so strictly at the reader's own risk. Because the use of information contained in this book is beyond the control of the author or publisher, liability for use is expressly disclaimed.

Chapter

1

Safety Systems

Safety systems include seat mounting, harness and belt mounting, proper driver apparel, proper car construction, and fuel cell use and mounting. Keep in mind that good safety systems are very important. Don't cut corners or your budget here.

Safety doesn't make the car go faster. In fact, it's a boring subject to racers. But it is something every racer **MUST** pay attention to. A driver's survival depends on it. Racing is a dangerous activity and serious wrecks do occur on short tracks. Don't skip this chapter.

Whenever you are building any type of race car, safety should always be the primary concern. Anyone who has ever attended a motorsports event is acquainted with the dangers of auto racing. It would appear that the racer would be the most responsive to the needs of safety equipment, designs and procedures in his race car — but, in many cases, apparently not so.

Some drivers will try to save money by buying substandard personal safety equipment. But, it may cost them more money in the long run in the event of a fire when you consider the costs of medical bills, loss of work, etc. If a racer cannot afford good safety equipment, he cannot afford to race.

While most racing organizations have some rules relating to safety, clear rule book definitions and tech inspections are often lax or even nonexistent. Terms such as "approved" and "recommended" still appear in rule books when relating to fuel cells, driving suits or safety harnesses. The choices of equipment and ultimate responsibility lie with the driver.

This section presents basic safety information. The information is presented to stimulate the racer's thinking. It is the race car driver's responsibility to

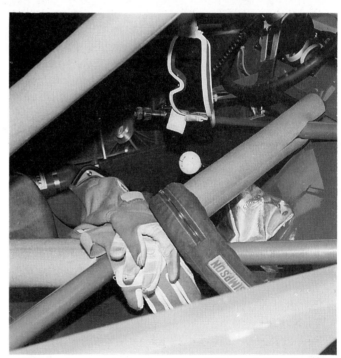

It is the race car driver's responsibility to fully educate himself on all safety equipment so he can make an informed purchasing decision. Every driver should make full use of all personal safety equipment available to him, regardless of which class of racing he is involved in.

fully educate himself from manufacturers' information on all safety equipment so he can make an informed decision about the equipment he purchases and uses.

Safety equipment is designed and manufactured to prevent injuries to the driver in the event of an accident. But safety goes much further than that. Safety, driver protection and crash worthiness are all

This piece of tubing (see arrow) is used to try to keep impact damage out of the rear crush zone (tubing frame behind it). With a hard horizontal push from behind, it can buckle under. It also prevents the front bumper of a tapping car from getting underneath the rear bumper of this car and lifting it upward to loosen the chassis.

very much a part of how you design and construct your race car.

Construction Techniques

Proper construction techniques means preventing mechanical failures. This means not only thorough preparation, but also utilizing good welding methods and techniques, utilizing the proper size materials, and the correct grade of metals. Joining, fastening, and even painting play an important role in keeping a race car together. Using the proper fasteners in the chassis adds greater strength and reliability. For reliability, you have to consider fastener ductility as well as ultimate strength. Make sure the nuts which hold the tie rod bolts, radius rod bolts, and all suspension attachment pieces are lock nuts (Nylock or fiberlastic stop nuts), or are safety wired, so that critical suspension and steering pieces don't come apart under racing conditions.

Using the wrong grade, type or size of fastener will lead to trouble. Magnafluxing or other forms of nondestructive testing should be performed on all critical parts of the steering and suspension on a regular basis. Firewalls should be as their name implies — fire proof. And don't leave holes in the firewall for fire to leak through.

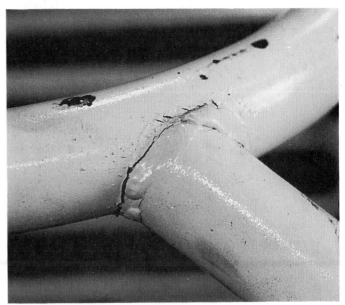

Using a light colored paint on the chassis and cage will quickly point out any stress cracks.

Notice how this well-prepared car has grommets around all tubes and wires that pass through the firewall. This effectively seals the driving compartment from the engine compartment, and prevents chafing of the wires and tubes.

The Firewall

A good firewall and sheet metal barrier will help prevent fire from entering the driver's compartment. The firewall is designed to keep fire, fumes and toxic materials out of the cockpit from the engine compartment. Occasionally there are holes left in the firewall from components that are removed. And, there is usually a gap around roll bars that penetrate

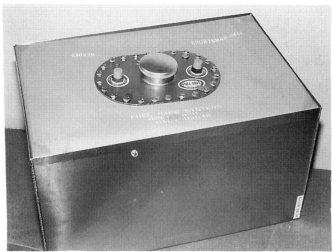

For fire protection, the best type of fuel cell to use is the type that uses a tough nylon rubber bladder inserted inside of a metal cannister.

There should always be a separate chassis tube and framework behind the cell which serves as a fuel cell guard. The bottom of it should extend to the bottom of the fuel cell.

through the firewall. These gaps and holes must be completely sealed up. Any opening left in the firewall is a hazard to the driver. A blown engine or a broken oil line can cause an under-the-hood fire which can blow through holes in the firewall. A broken radiator hose can blow boiling water back on the driver through a hole in the firewall. Seal the holes with metal plate coverings, and use a plate to seal gaps around roll bars or weld the gap shut.

Driver Safety and Attitude

The driver himself is often the most dangerous element on the track — both to himself and to others. Burning the midnight oil and expecting to drive at your best the next day is not very sensible. When tired, the mind reacts more slowly and judgment is impaired.

This same type of warning should go for the angry or revengeful driver. Short track racing at times can be a contact sport, and some guys like the contact more than the sport. Being the "contacted" driver can be infuriating, but driving to get even with another driver can only distract and lead to trouble.

An additional safety factor is to remember to place jack stands underneath the race car when it is elevated with a jack, no matter if the car is at the shop or the race track. So many times racers get in a hurry to fix something under the car and they forget about putting a solid support under it. If the car falls off the jack, severe injuries can result.

Using A Fuel Cell

A fuel with a metal canister is mandatory. Make sure it is a true fuel cell, not just a plastic gas tank. A true fuel cell is made of a special tough material that will deform without splitting open. It also contains a special foam inside to minimize the movement or sloshing of fuel and the build-up of air pockets. A fuel cell should also be mounted inside of a properly designed metal canister (constructed of 22-gauge steel) to give extra support to the cell in case of an impact to it.

There are two different styles of fuel cells available. The top of the line style of cell uses a tough nylon rubber bladder – inserted inside of a metal canister – which performs well under high impact situations. The other style of cell is generally sold as a "sportsman, or budget" style, which means it is a less expensive cell. It uses a polyethylene fuel cell held inside of a steel canister. These, generally, are not recommended for high impact situations because they do not have the elasticity of the nylon rubber bladders. Extra care must be taken when using a budget style of cell so that it is more protected by the car's chassis members.

Any fuel cell should be mounted as far forward in the chassis as possible, with the bottom of it no lower than the bottom of the rear end housing tubes for

A common method of mounting a fuel cell is on a tube frame base made from 1-inch square .095-inch wall tubing, and securing the cell using 1-inch wide, 1/8-inch thick steel strap. Be sure to check if your track's rules vary from this.

protection of the cell. Additionally, there should be a separate chassis tube (minimum 1.5-inch O.D.) mounted behind the cell which serves as a fuel cell guard. This guard must extend to the bottom of the fuel cell. Note that some racing associations have specific rules for the mounting height and location of the fuel cell, so check with your rule book.

Fuel Cell Mounting

The fuel cell canister must be properly installed into the chassis to prevent it from being torn loose in a severe impact. You can use 1-inch wide, 1/8-inch thick steel strap to secure the cell. Set this on a tube frame base made from 1-inch square, .095-inch wall tubing. To resist lateral movement, use 2-inch "L" angle metal brackets, 1/8-inch thick, on each side. But check with your local tech inspector. Specifications or requirements may vary.

Remember that the cell must be properly vented with a steel check valve in case the car gets upside down. The fuel cell cap should also be tested periodically to make sure it does not leak when the cell is upside down.

Driver Restraint

Restraining the driver's body will prevent injuries from contacting the interior of the car, a wall, another car or its parts, or even the track surface. The restraint system consists of the lap belts, shoulder harnesses, the crotch belt, neck collar, helmet re-

Enclosing the left side door bar area with steel sheeting provides an extra level of penetration prevention for the driver. Many racing associations are now making this mandatory.

straint strap, and the window net. The restraint system also helps prevent injuries from the inertia generated by a flip or crash.

Are you thinking here that you race at just a short track where the speeds aren't too high, and the safety equipment we are specifying really doesn't have to apply to you? You are wrong. Take this seriously. Accidents can happen even at the slowest of speeds. If you follow the proper safety procedures and use the proper equipment (and always buy new, not used, equipment) your chances of serious injury can be very much reduced.

Driver Impact Prevention

The best way to prevent the driver from making contact with the interior of the car is to take the driver's height and size into consideration when positioning the seat. A general rule of thumb is to allow a minimum of four inches between the top of the roll bars and the top of the driver's helmet.

The rest of the roll cage should be located as far as possible from the driver's head and arms. Remember that under severe impact the neck and spine can stretch as much as ten percent, and a belt can stretch as much as two inches per foot. All that combined can cause the driver to reach out of the window or

The best type of roll bar padding is the high density foam. Using the type with the offset hole, which places a 1-inch thickness toward the driver, adds more driver protection.

over to the passenger side. Or, worse yet, the driver can contact the track surface if the car rolls over or is upside down.

Contact with the interior of the car can be cushioned with a high density foam padding, such as the Longacre ProTecto or Reb-Co Pro Bar XLT, to improve shock absorption. This type of padding is a very dense foam which results in better cushioning and rebound ability. This padding is made with an offset mounting hole so one inch of pad cushioning is available to mount toward the driver for more

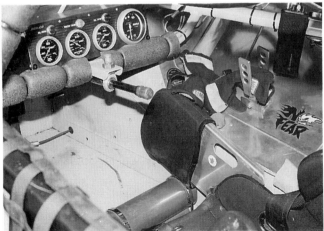

Make sure all bars that the driver could possibly come in contact with are properly covered with high density roll bar padding, including padding around the steering column to protect the driver's knees.

protection. To attach the roll bar padding, use 3M Weatherstrip Adhesive in combination with tie wraps in critical areas to hold the offset in the proper position.

Any type of roll bar padding used in a race car should be flame retardant so that the material will not melt and drip on a driver in case of a fire. Molten plastic foam can cause extremely serious burns. Always use a roll bar padding made specifically for race car use. Never use air conditioning line foam wrap – it won't provide and type of impact protection.

Make sure all bars that the driver could possibly come in contact with are properly covered with high density roll bar padding. A good idea would be to add padding around the steering column to protect the driver's knees, and around the gear shift lever also.

Window Nets

A window net will help keep the driver's arms and head in the car should his body stretch that far, and the net will also prevent debris from hitting the driver. The window net attachment to the car should also be carefully considered. NASCAR requires the use of a passenger car seat belt attachment clamp to connect to the upper window net retaining rod at the top and front as a quick release mechanism.

Make sure the window net fits snugly when it is in place so that stretch in the material is tight enough to prevent the driver's head or arms from going outside the window opening plane during an impact. And, make sure that all clamps and brackets which hold the window net in place will sustain a heavy blow and still keep the net in place. To insure net strength, it should be sewn with Kevlar thread.

It is extremely important that the window net be very easy for the driver to open and remove from inside the car all by himself, even if there has been damage to the driver's side of the car, or the car is upside down. Be sure your driver can do it. The ease of opening the net can save a driver's life. Take this seriously. If the car is surrounded by flames, it may be awhile before somebody rushes in to release the window net for the driver.

Most track's rules require a positive quick release mounting of the window net. The upper retaining rod for the net is held at the rear in a spring loaded bracket, and at the front by a seat belt latch which can easily be reached by the driver from inside the car.

Other Cockpit Items

The ignition components, fire extinguisher, and other attachable items must be located where the driver cannot possibly contact them under a hard

An electrical kill switch, or master disconnect, shuts off all electrical systems in the car in case of an emergency. This one is mounted beside the driver where it is easy for him to reach, but much more difficult for outside emergency workers to find. It is mounted so that the negative side of the battery is disconnected by the switch.

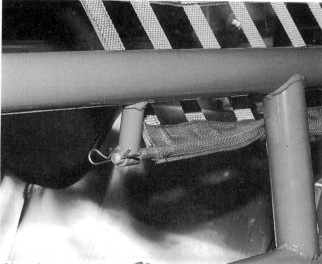

Note the sturdy mount for the lower window net retaining rod. It is held in place with a hood pin which the driver can easily pull out. Make sure window nets are easy to remove from the inside by the driver.

impact, or where the driver can get caught up on them during a quick exit out of the car.

Quick release steering wheels are a requirement to aid fast ingress and egress. But, just in case the quick release mechanism is jammed by impact from a wreck, make sure the driver can get out from under the steering wheel if it cannot be released.

It is mandatory in many association's rules to use a master electrical kill switch which shuts off all electrical systems. The cut-off switch must be installed within the reach of the driver to facilitate complete shut-down of the electrical system in case of an emergency. The shut-off switch should be clearly marked to be visible from the outside of the car, so it is readily accessible to rescue workers in case of an emergency. The shut-off switch also comes in handy when working on the car's electrical system and to shut the whole system off when the car is in storage.

Escaping a crashed race car can be time consuming, especially if a driver is stunned or injured. The safety minded driver should practice a logical sequence of escape from his car, and store the information in the back of his mind. Frequent mental imagery of this procedure will program the driver's mind to react to an emergency, so even under the most extreme emergency the driver will be able to react without even thinking about what needs to be done.

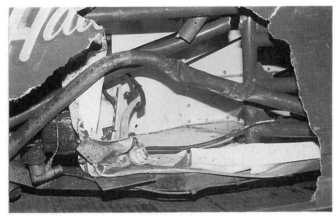

The leg brace is one of the most important tubes in the chassis. It protrudes outward over the left front frame corner to protect the driver's feet and legs.

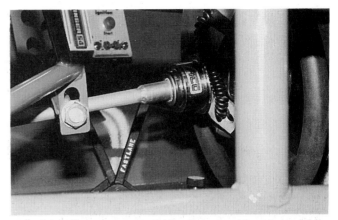

A quick-release steering hub allows a driver to quickly get the steering wheel out of the way to exit the car. This one uses a splined coupler which has a spring-loaded flange at the rear of the hub which simply pulls forward to release the steering wheel.

The driver has the responsibility every time he enters his race car to survey every inch of the cockpit. Even though he is familiar with it, he should look for anything that has been changed or loosened, tools that have been left inside, panels that are loose, etc. The driver cannot ever take anything for granted or become complacent.

Quick-Release Steering Hub

Crawling in and out of a stock car is always a difficult situation. But get the steering wheel out of the way, and it makes it a lot easier. And when the car has crashed and you need to get out of the car in a hurry, getting the steering wheel out of the way quickly helps out a great deal.

There are a lot of different quick release hub designs on the market. Research the differences and find one that best fits your situation.

Some quick-release hubs can be released in any position, whereas some require you to pull a pin in one position. Some of the hubs fit easily to the shaft, while others are more difficult to fit. Some use a fine spline mating system, while some use a course spline. And, some use a hexagon mating system. While the hexagon is easier to install in a hurry, it tends to get sloppier in fit as they wear than splined hubs.

Pinless quick release hubs have a spring-loaded flange at the rear of the hub which simply pulls forward to release the steering wheel. This makes it so much easier than having to search for a release pin, especially in a panic or rescue situation.

Be sure that your quick-release hub does not contain any plastic parts. Plastic can melt in the event of a fire, making the steering wheel difficult or impossible to get off. When an all-metal hub is used, be sure to periodically check all parts for signs of corrosion. Always make sure the safety latch on the hub is properly seated when the steering wheel is installed. And be sure the safety button is properly lubricated so it works freely.

Fire Suppression Systems

Every car should at least have a properly charged and certified fire extinguisher securely mounted in the car. A good place to mount the bottle is along the driveshaft tunnel, within reach of the driver when he is strapped in. An alternate position is just forward of the driver's seat.

Be sure that the extinguisher is securely mounted. If it isn't, it could become a deadly missile in the event of an accident or rollover. It could seriously injure the driver.

On-board fire suppression systems can help put out a fire in a hurry, or at least allow the driver some critical extra time to abandon the car in case of a fire. This system uses a fire extinguisher connected to remote lines and nozzles that direct the fire fighting agent to the engine compartment, the driver's compartment and the fuel cell area.

Most race car fire extinguishers have used Halon 1211 or 1301 as the fire suppression agent. It is a colorless, odorless nonconductive gas which does not leave a residue to clean up. It also does not block a driver's vision when it is discharged. Halon provides about 2 seconds of protection for each pound

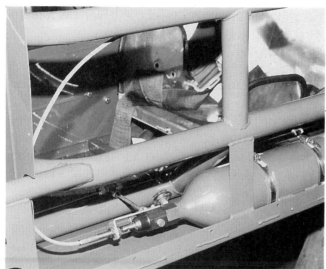

On-board fire suppression systems can help put out a fire in a hurry, or at least allow the driver some critical extra time to abandon the car in case of a fire. Make sure that the fire bottle is securely mounted.

of agent in a bottle in a 3-nozzle system. Bottles are available generally in 2.5, 5, 7 and 10-pound sizes. Fire suppression systems are installed to save the driver and give him ample time to exit a burning car. They are not designed to save an entire car in the event of a raging fire.

Production of Halon 1211 and 1301 has ceased because of its negative implications on the Earth's ozone layer. However, there are millions of pounds of this material in existence, and it still can be used in race car fire suppression systems. It is the only fire extinguishing product that is FAA approved for commercial airliners. It is approved by the EPA for use in occupied areas, meaning it is not highly toxic, and it is safe to breathe without ill effects for a limited period of time.

There are several different compounds being developed to replace Halon. One of these is CEA-614 from 3M. It has been satisfactorily tested by the Air Force. Additionally, Haloton 1 from American Pacific Corp. is being used as a Halon 1211 replacement in race car fire systems. It is approved by the EPA, UL, and FAA.

Be aware that whatever compound ends up being used as a replacement, the nozzles used with a Halon 1211 or 1301 system will have to be replaced. Different chemicals have different densities, so they will require different nozzles and seals.

Racing Seats

A top quality racing seat that properly restrains the driver during cornering (or in a wreck) is extremely important. Be sure the driver is comfortable and properly restrained in the cockpit. The seat should fit so that it and the tightened safety belts hold the driver in comfortably while cornering at the car's maximum G force. It will wear a driver out trying to hold himself in position in a poorly fitting seat by tightly grasping the steering wheel. The arms and shoulders should be relaxed so the driver can concentrate on driving.

Take the opportunity to explore the differences between racing seats. There are big differences. A proper seating environment is essential to help a driver perform at his highest level. The optimum seating position and body support enhances driver efficiency by increasing upper body strength, promoting proper respiration and blood circulation, protects the spine, and spreads the force of an impact over the maximum area of the body. It also gives proper support to the torso and head to minimize whiplash injuries.

The seat has to be strong enough to resist crushing under an impact and several secondary impacts. Any seat that deforms or fractures under impact leaves a driver totally unsupported and vulnerable to injury.

For that reason, aluminum seats are far superior to fiberglass or plastic seats. The racing seat must be capable of supporting the driver's body in the event of a high G-load crash. The seat material thickness should be a minimum of .125-inch. Anything less than that doesn't offer adequate strength. Many of the top-of-the-line and custom seats use two layers of .090-inch aluminum.

A proper racing seat should be high-backed seat. The high back on the seat provides a head support instead of using a separate head rest. The seat should provide an even amount of support to the thighs, hips, torso and shoulders on both sides of the driver. A seat with only a simple rib support under the driver's arm pit can cause the driver heavily bruised or broken ribs in the event of a heavy side impact. Bolt-on head supports on both sides of the seat will provide head and neck support for a driver in the event of a side impact.

A good racing seat will have a slot in the forward area of the seat bottom for the anti-submarine belt. That is because this belt is an absolute must to use

A good racing seat should comfortably provide even support on the right side of the torso and shoulders. Racing seat attachments -- lumbar support, head and shoulder rests, leg supports, and double rib cage wrapping -- help support the driver's body and increase safety.

in conjunction with the rest of the belt and harness system.

Every seat must have some type of multiple bend or aluminum channel section extending across the front of the seat bottom to add strength. This prevents seat buckling or collapse in the event of a major impact and helps keep rigidity in the seat.

Leg support extensions on the seat are also important to protect the driver. A heavy side impact can whip a driver's legs around in an unsupported seat. This, or the legs striking something in the car, can cause leg fractures, or knee or ankle injuries. The seat extensions keep the driver's legs supported and in place.

The back of the seat should be slotted just at or slightly above the driver's shoulders to allow the shoulder harnesses to pass through. Be sure that the belt slot edges are rubber or plastic lined so that the sharp edges of the seat will not chafe or cut the belts.

Racing seats are made in a variety of sizes and configurations, and are also custom built for a par-ticular driver as well. Contact a seat manufacturer to be sure you get one that fits you properly.

The seat should be fully upholstered, and provide a snug fit (but not tight) for the driver. Seat padding helps to absorb vibration and impact loads, which makes a driver more comfortable. This also helps a driver concentrate better. Padding should be a very dense foam, but not thick. When purchasing seat covering and cushioning material, be sure it is flame retardant and will not melt in the event of a fire.

Seat Mounting

The racing seat should be mounted to the chassis via a steel tubing seat hoop. The hoop should be constructed of 1.25-inch O.D., 0.125-inch wall round steel tubing, and should be mounted to the left on the chassis frame and to the roll cage behind the seat. The seat hoop should also be welded in a third location near the right side to prevent pulling up in case of an accident. Such a design will cause the seat to move with the chassis in the event of a hard impact to the left side of the car. Any plates or brackets welded to the seat hoop should be a mini-mum of 3/16-inch thick, and care should be exer-cised not to drill mounting holes too close to the edge of the bracket to prevent fatigue cracking from the hole.

With the seat, use 3/8-inch O.D., 0.065-inch wall round tubing to form a supporting frame around the complete perimeter of the seat, across the back and down the center of it. This frame gives secondary support for the driver should an impact ever break the seat.

The seat must be mounted at six positions on the seat hoop — four on the bottom of the seat and two on the back of the seat. Be sure to use rounded head bolts to insure driver comfort when mounting the seat, and use large diameter fender washers under each bolt head to prevent the bolts from pulling through the seat, or work-fatiguing the seat. Use 3/8-inch bolts with lock nuts.

Seat mounting bolt tightness should be checked on a regular basis. And when doing so, be sure to check the weld integrity on the seat mounting brackets.

Remember that the basic function of the seat is to maintain the driver in a certain predetermined posi-tion during a race that is comfortable to the driver. And, in case of an accident, it must be strong enough to support the driver and keep him restrained in position. Should the seat deform or break up, the driver will no longer be held in place by the safety belts and harnesses.

Driver Protective Clothing

The driver's firesuit is one of the most important safety items he has. Fireproof underwear, gloves and shoes are also important to use to prevent painful burns. In the event of a fire the driver will want to escape promptly. The escape would be very difficult if the driver's hands and/or feet are burnt.

NEVER drive a race car without some form of a fire and heat retardant driving suit. There are several different types and qualities of material used for driving suits, and the driver should research the various types and make himself familiar with all of them before making a purchase. The different types and layers of fire retardant fabric will provide varying amounts of thermal protection to the driver. Multi-layer driving suits — preferably three-layer suits — are highly recommended to give the driver as much fire protection as possible.

In a fire, the driver is faced with two major problems — flame and heat. The basic firesuit will help guard against flame, but the intense heat requires much more protection.

One of the most important jobs of a firesuit is to prevent or retard the transfer of heat to the driver's skin. A single layer suit will do the worst job of this. A 2-layer suit will generally double the protection (exact time depends on the type of fabric used). It does this not only by doubling the amount of insulating fabric, but also by having air trapped between the layers. The trapped air functions as an additional insulator. So, you can see that a 3-layer suit will add even more thermal protection.

A very minimum for driver safety should be a 2-layer suit used along with fire retardant Nomex® or PBI underwear. In severe heat or flames, any type of flame retardant fabric is going to degrade, char and disintegrate. If a driver is in a fire where it takes him a while to get out of the car (such as having to stop the burning vehicle from a high speed), the multiple layers become extremely important in protecting the driver.

When considering the purchase of a driving suit, price should not be a determining factor. Don't buy the cheapest driving suit that meets the minimum

This 3-layer driving suit is cool even in hot weather, and it provides excellent driver protection. A good helmet, gloves, fire retardant shoes and underwear are good insurance. Gloves should be two-layer Nomex® on the glove back and one on the palm, and should fit tightly.

rules of your racing association. Spend the extra money for extra protection.

The driving suit you purchase should have an SFI rating label on it. SFI is an independent testing organization that establishes and monitors minimum performance specifications for racing safety products. Minimum standards of performance are established to protect the consumer. Any SFI-labeled product is tested annually to insure continued product quality.

The SFI rating system assigns a number to a product based on how much protection it provides, using tests with heat and direct flame. The least protection is rated SFI-1, while the greatest protection is rated SFI-20. A driving suit rated SFI-1 gives only three seconds of protection, while SFI-5 gives nine seconds and SFI-15 gives 30 seconds. This shows that you have to purchase a suit with substantial protection and layering to give yourself just a minimal amount of extra time to exit a fire.

Also, remember that the stated protection times have been established in a laboratory by SFI. All conditions in the lab were carefully controlled, and the driving suit material was new and clean. When the suit fabric ages, or gets embedded with grease, oil or fuel, the protection time is going to be less.

Driving suits should always be stitched with Nomex® thread, never nylon or polyester. And don't ever use a suit which has a plastic zipper. All of these materials will melt in a fire. SFI-rated suits always use Nomex stitching and brass zippers with Nomex® backing. Buy a driving suit from a major brand name manufacturer to insure the quality of construction. And, make sure that the driving suit does not fit too tight. A small air gap between the suit and the driver's fire retardant underwear can add a small amount of extra protection because the air gap requires a longer time to transfer heat to the next layer.

Another consideration for racers is that higher quality driving suits are available in many colors, which can serve to enhance the image of the racing team. Suits are now available in black, blue, red, yellow, and other colors, as well as stripes and multiple colors. Many drivers are getting away from white because it shows dirt too easily.

Treat a driving suit like an expensive piece of fine clothing. To protect the fabric of a driving suit, it should be regularly cleaned. Most driving suit makers recommend that a suit be washed in cool water or be dry cleaned. Do not wash in hot water, or dry in a hot dryer. Follow the cleaning instructions that come with the suit.

Fire retardant underwear (socks, Long Johns, long sleeve tops and face hoods) will add several seconds of safety if the driver is exposed to fire. A driver can have a uniform that can withstand fire for two days, but it will still transfer heat to the skin underneath it. Areas of the body such as the neck, ankles and wrists are often burnt when the driver does not wear fire retardant undergarments. Full underwear is highly recommended to drivers using only a single layer driving suit. However, a multi-layer suit and underwear will greatly enhance the chance of escaping injury from fire and heat. Underwear made from PBI fabric rather than Nomex will be cooler because PBI absorbs and transfers moisture, which means the driver will remain cooler.

Gloves are important in a fire to allow the driver to remove his equipment before bailing out without burning his hands. If the driver burns his hands, he will not be able to release his harness or door latch, or even get out of the car. Gloves should be two layer Nomex on the back and single layer Nomex on the palm side with one or two layers of leather for positive grip and control. Gloves should fit tightly. They should not have any leather that touches the skin.

Flame retardant shoes are equally important. The driver will have a hard time escaping a burning car on blistered or burnt feet. Specially designed racing shoes from safety apparel manufacturers offer two advantages. One, they are flame retardant. And two, they are pliable and have a thin sole so that the driver can feel the throttle a lot better. So much of racing is fine-line throttle control. Heavy and stiff boots or leather shoes won't get the job done here. Thin, pliable ones make a big difference in control and feel.

It is highly recommended that the driver take advantage of all the protective clothing that is available. It can save his skin.

Belts and Harnesses

The driver's restraint system (belts and harnesses) is probably the most important safety system in any race car. Belts and harnesses must absorb a tremendous amount of load in accidents. To be sure of getting the best for your money, only purchase belts from reputable U.S. manufacturers, such as Pyrotect, Simpson or Deist. Some foreign harnesses are plenty strong, however the buckles will cut the fabric before the maximum allowable forces are encountered.

One of the most common mistakes with harnesses on stock cars is that the installers use much too long of a shoulder harness. A 3-inch belt can stretch as much as two inches per foot in a good size crash. Head tilt, neck stretch, slack in the harnesses and two inches per foot of belt stretch can easily allow a driver to hit the steering wheel, or worse, the driver can slip out of the harnesses altogether.

Another thing to avoid is installing the shoulder harness mounts too far apart from each other. This will cause the harnesses to slip off the driver's shoulders during a race. The double strap type of shoulder harness (each strap having its own mount instead of being sewn together in a "Y") is much stronger and should be used. The double harnesses should mount behind the driver from four to six inches apart with a short length of strap behind the driver to prevent unnecessary strap stretch. The shoulder harnesses should be mounted on a cross tube right behind the

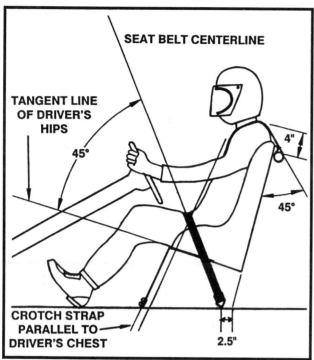

SEAT BELT CENTERLINE

TANGENT LINE
OF DRIVER'S
HIPS

45°

4"

45°

CROTCH STRAP
PARALLEL TO
DRIVER'S CHEST

2.5"

The correct mounting locations and angles for the driver restraint system.

These shoulder straps are mounted way too low. They should be mounted at a level about 4 inches below the height of the driver's shoulders in order to prevent compression of the spine in a heavy impact. It also appears the driver is routing the shoulder straps around the seat back instead of through the slot. This is a bad practice because it will allow the straps to slip off the driver's shoulders, and he cannot get them tight enough this way.

seat at a level about 4 inches below the height of the driver's shoulders in order to prevent compression of the spine in a heavy impact. Mounting the shoulder harness this way will diminish the amount of belt stretch encountered during an impact. The shoulder belts should be mounted to the chassis so there is always a straight pull on them from the mounting point to the driver's shoulders under tension. Always use 3-inch wide shoulder harness straps in order to spread the forces of a crash over a wider surface area of the driver's body.

The nylon webbing of belts and harnesses are susceptible to ultraviolet radiation. In other words, exposure to sunlight helps to wear them out. It will reduce the tensile strength of the material over time. And, belts are always subjected to dirt and oil which, over time, embeds into the nylon material causing material abrasion of it. This further weakens the tensile strength of the material.

More and more major sanctioning bodies are requiring the use of date stamped safety belts and harnesses. These are becoming very standard. All belts and harnesses should be replaced after two years, even if they are not damaged. Many top racers replace theirs every year.

You must be aware that if you buy a used race car, or if you buy used belts, that they may be close to or beyond their expected life. And that plays a big part in your safety. The best thing to do is always start out with new safety belts and harnesses so that you can track age, use and abuse of them yourself. And, make sure that any belt or harness you purchase has an SFI certification tag attached to it.

When the belts get old and worn, you can send them back to their manufacturer and have them rewebbed. The manufacturer will sew on new belt webbing material using your hardware, and the cost is about half that of a set of new belts.

The correct positioning of the belts on the driver's body is vitally important to protect him from injuries. The lap belt should go around the hips – never the stomach – and mount at a rearward angle of 45° from the flat of the driver's seated position (see drawing).

The anti-submarine or crotch belt is used to prevent the driver from sliding out from under the lap

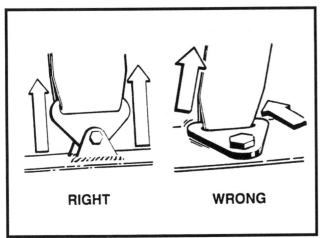

RIGHT   WRONG

The mounting brackets for belts and harnesses must be installed at an angle compatible with the direction of pull on the webbing under full load. Drawing courtesy of Simpson.

belt in the case of a heavy frontal impact. This belt also prevents the lap belt from riding up into the abdominal area and helps keep tension on the shoulder straps. The crotch strap should be mounted at a forward angle from the seat which places the belt parallel to the driver's chest.

Seat belts and shoulder harnesses should be as tight as the driver can possibly stand them, and the driver should always retighten them when he is on the track as often as he can, such as during caution laps.

Helmets

The helmet is a very important piece of safety equipment. Don't take any shortcuts here. Buy only the best. If your budget won't allow it, don't race until you can afford to buy an excellent one. Be sure to purchase a helmet that bears the latest "Snell" sticker and you will comply with all the latest safety requirements for helmets. The current Snell standard is SA 95 for helmets with a Nomex interior, and Snell 95 for helmets with a nylon interior.

Follow these tips with your helmets:

1) When purchasing a helmet, make sure it fits snugly, but not too tight, as headaches may result. Keep in mind that your head will swell slightly on a hot day. The helmet should not be so loose with the chin strap off that it can rotate on your head while you try to twist it.

2) When adjusting your helmet before going out on the track, always tighten the strap as tight as you can stand it and press down on the top of the helmet

Routing the shoulder harnesses through the slot in the seat back keeps them properly located so they will not slip off the driver's shoulders during an impact.

to settle it on your head. Just like with seat belts, the chin strap will loosen up in a few minutes from the helmet settling. Make sure it is tightened up.

3) With an open face helmet, use goggles or a shield to protect your eyes from dust and flying debris.

4) Use caution when using a full face helmet and a shield. In an enclosed car, the driver may not get enough ventilation to replenish the oxygen supply inside the helmet. This situation is worsened by the use of a neck brace or head sock. There are full face helmets designed for use in a stock car that feature increased venting, air induction, and circulation. Be sure that any full face helmet you purchase provides you with good visibility. A full face helmet can offer additional protection to the driver's face in the event of an impact. Helmets constructed out of Kevlar or carbon fiber are lighter in weight, but still offer the same amount of protection as a fiberglass helmet.

4) Do not sand or drill on a helmet. Some paints will attack the fiberglass, while sanding and drilling will weaken the structure. Use only acrylic enamel paint, never a lacquer. Lacquer paints will weaken the shell.

5) Remember that helmets are designed to absorb an impact only **one** time. Any time that a helmet has absorbed an impact – even if there is no apparent damage – it should be returned to the manufacturer for inspection before ever being used again.

Chapter

2

Front Suspension & Steering

Front Suspension Layout

There are a wide number of topics to consider when designing the front suspension geometry layout — spindle choice and its dimensions, kingpin and steering axis inclination, wheel offset, frame height and clearance (which influences lower pivot point heights), car track width, camber change curve, static roll center height, and roll axis location.

The major areas of concern in front suspension design are bump steer, camber gain and roll center. The most important parameter to establish first of all is the roll center height and lateral location. The roll center is established by fixed pivot points and angles of the A-arms. These pivot points and angles also establish the camber gain and bump steer.

Front Roll Center

The roll center height determines what percentage of the overturning moment (inside to outside weight transfer) will be distributed onto the tire contact patch as downforce, and what percentage is received as lateral loading against the tire's tread face. Vertical loading creates downforce on the outside tire, so the more vertical loading there is, the better the outside tire sticks during cornering. This downward tire loading is why traction increases as track banking angle is increased. A lower front roll center will create more vertical loading on the outside tire contact patch. A higher roll center will load the transferred weight more horizontally, which creates a shear force at the tire contact patch. A car needs body roll during cornering to transfer weight down-

ward onto the outside tire contact patches. This is a product of a lower front roll center. If weight was transferred laterally to the contact patches, the tire rubber would shear across the track surface and the car would slide out.

The front roll center of a paved track stock car should be located between 1.5 and 2.5 inches above the ground, and offset 3 inches to the right of the vehicle center line. The upper and lower control arms should be placed so that the instant center which their mounting points form is located 1 to 2 inches inside of the opposite lower ball joint.

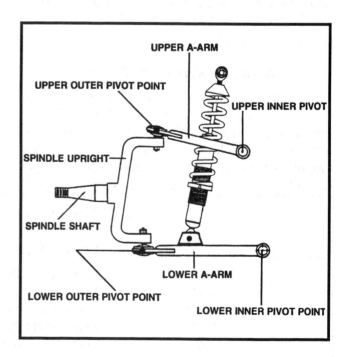

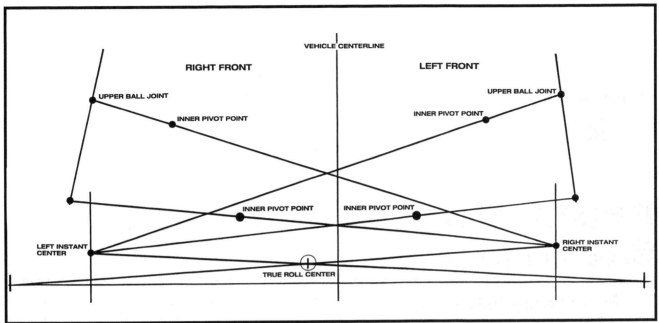

How the front roll center is determined: First, lines are projected through the pivot points of the upper and lower control arms of the right front until they intersect. This intersection is called the instant center. Then a line is drawn from the instant center to the right front tire centerline. This line is called the swing arm length. Then this same process is applied to the left front suspension. Where the swing arms for the left front and right front intersect is the true front roll center. The front roll center of a paved track stock car should be located between 1.5 and 2.5 inches above the ground, and offset 3 inches to the right of the vehicle center line.

See the adjoining table for roll center height recommendations for various types of applications.

To achieve the front roll center height required for paved track cars with a fabricated front clip, a 7.75-inch tall spindle with a 10° kingpin inclination is generally used. Cars which have to use stock type spindles will have to use a slightly taller spindle, such as 8.75 inches.

Instant Center Width

The instant center width controls how the roll center behaves during body roll. The wider the instant center width, the less negative camber gain achieved at the right front during body roll. A narrow instant center width creates a more radical change. A shorter instant center width also makes the roll center height move up and down more radically during body roll, which has a major effect on the loading of the right front tire contact patch at corner entry and mid-turn. The recommendations presented here for roll center height and instant center width take all of these elements into consideration. They present the ideal instant center width that doesn't make radical changes to the suspension geometry during body roll, which keeps the roll center location very stable.

The correct instant center width depends upon the track width of the race car. A 66 to 68-inch front track width requires an instant center that is located 2 inches inside of the opposite ball joint. With a 63 or 64-inch track width, the instant center should be located 1-inch inside of the opposite ball joint. With

Front Roll Center Height Guideline

Roll Center	Chassis Type	Motor Type	Track Banking
2.5"	Stock front clip	Cast iron heads	0° – 12°
2.125"	Stock front clip	Cast iron heads	13° – 18°
1.875"	Fabricated clip	Aluminum heads	0° – 12°
1.5"	Fabricated clip	Aluminum heads	13° – 18°
2.5"	I.M.C.A. Modified	Cast iron heads	0° – 12°

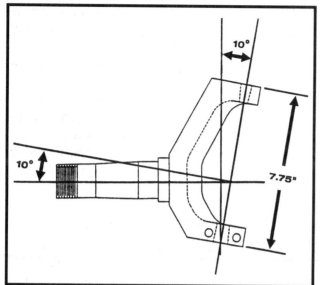

To achieve the front roll center height required for paved track cars with a fabricated front clip, a 7.75-inch tall spindle with a 10° kingpin inclination is generally used.

The Coleman variable length upper ball joint can help to locate the upper outer mounting point higher above the spindle upright to adjust roll center height. There are two spacers -- 3/8-inch and 1/4-inch -- for a total of 5/8-inch difference in spacing that can be used.

a narrow front track width, such as ASA cars, the instant center should be located right at the opposite ball joint plane.

The True Front Roll Center

The true front roll center of the independent front suspension is that point where the instant center swing arms of the left front and right front suspensions cross each other. The height (1.5 to 2.5 inches) and the offset (3 inches) that we referred to previously is the true roll center at this point.

Roll Center Offset

It is important to locate the front roll center 3 inches to the right of the vehicle center line. This creates the leverage required as the body rolls to get enough downward force on the right front tire to get it to stick. This means slightly more than half of the left side sprung mass is rotating about the roll center. This allows the right front to steer and turn properly in the middle of the corner. However, if the roll center is placed too far to the right of vehicle center line, there will be too much leverage lifting up on left front as the body rolls. This will load the right rear tire heavily under acceleration. While this might make the car fast for a few laps because of heavily loading the right rear, it will quickly overheat and

wear that tire, and then it will not give any traction at all.

If the roll center is located to the left of the vehicle center line, the car will not turn into the corners well. There is not enough leverage generated with body roll to stick the right front tire properly. The right front tire will slip sideways at mid-corner. Additionally, under acceleration off the corners there will not be enough lift from the left front onto the right rear to add traction to that tire. A car with the roll center located to the left of vehicle center line will typically push going in and be loose coming off the turns.

Other Influencing Factors

While a paved track stock car's front roll center should fall, in general, between 1.5 and 2.5 inches, there are specific heights applicable to specific types of cars. The correct roll center height depends upon several influencing factors. These include the height of the major weight masses in the front of the car, the banking angle of the track, and the type of tire being used.

A car with a higher center of gravity or higher front weight mass wants to roll over more during cornering as compared to a lower weight mass car. To control the extra body roll, the front roll center height has to be raised to add more stability to the car. This

Much more of the cornering force is experienced as down force rather than side force on a medium or higher banked track. A car on a banked track doesn't get a lot of body roll, so it doesn't need a lot of camber change. The front roll center should be lower so that less negative camber gain is produced at the right front.

is more preferable than using extra stiff front springs to control the excessive body roll.

A major contributing factor for front weight mass is the type and placement of the motor. Mass placement in the front end has a big influence on roll center height. Less or lower front mass requires a lower roll center. A larger or higher front mass requires a higher front roll center. A motor with cast iron heads weighs more than a motor with aluminum heads. The roll center height depends on how high the motor is mounted in the chassis and how far the setback is.

If an aluminum head motor is used and placed low in the chassis, the front roll center has to be lower to get weight to transfer to the outside tires during cornering because of the lower weight mass. If a cast iron head motor is used, that is an additional 40 pounds of mass placed 20 inches higher in the chassis. This combination would use a slightly higher front roll center because of the higher weight mass.

The frame and suspension contribute to front weight mass as well. A stock type of front frame clip weighs more than a fabricated front clip, so a car with a stock frame will require a slightly higher front roll center.

The type of tire used on the race car is also an influencing factor. If a hard compound spec tire is used, a lower front roll center is required to create more downforce on the outside tires. A lower roll center creates more body roll and, combined with lighter spring rates, provides the traction and side bite that hard tires require.

The most critical element is to have the best balance between mass placement and roll center location so that the car turns in the middle of the corners. This happens when sufficient weight is transferred to the outside tires to create vertical downforce on them.

Lateral Tire Scrub

A very important byproduct of a lower front roll center is that it yields a very minimal track width change during body roll, and thus minimal lateral tire displacement or scrub across the track surface. This is very important because the lateral scrub during bump travel puts extra heat into the tire tread and cords. With a higher front roll center, the front tires will perform well for maybe 10 to 15 laps, and then they will start to go away due to excessive heat build-up. A lower front roll center prevents this effect.

Effect of Banking Angle On Roll Center

The amount of banking of the race track's corners has an influence on the front roll center height. With a medium or higher banked track, much more of the cornering force is experienced as down force rather than side force. This means a car on a track that is banked 15° or more doesn't get a lot of body roll, so it doesn't need a lot of camber change in the suspension. For higher banking angle applications, the front roll center should be lower so that less negative camber gain is produced at the right front.

The camber gain curve of a front suspension is directly tied to the roll center height. A high roll center creates more negative camber gain at the right front during body roll. The LOWER the front roll center, the LESS camber gain per inch of wheel bump travel. The HIGHER the front roll center, the MORE camber gain per inch of wheel travel.

Jacking Effect

Another factor to consider when designing the front roll center height is the jacking effect it pro-

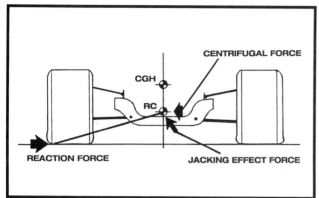

Jacking effect is a reaction from the outside tire back to the roll center, which places a force pushing upward on the roll center.

duces on the chassis. The jacking effect is a reaction from the outside tire back to the roll center, which places a force pushing upward on the roll center. This jacking effect could best be graphically described as the same type of force which pushes a pole vaulter up over the bar after he has planted the end of this pole. In this example, the pole vaulter is the roll center, and the planted end of the pole is the outside edge of the outside tire. The forward momentum of the vaulter is comparable to the centrifugal force acting on the car's body during cornering. In the vaulting example, the greater the forward momentum and the greater the height of the vaulter, the greater the force vaulting him over the bar. Applying this in terms of the roll center, the higher the roll center and the greater the centrifugal force, the more reactive force there is pushing upward on the roll center, which in turn actually raises the physical height of the race car.

A lower front roll center helps to minimize the jacking effect. If the front roll center is placed at ground level, there is no jacking effect. As the roll center rises above ground, the forces encountered through the jacking effect increase.

Designing The Front Suspension Mounting Points

Knowing the basic variables required for the correct front suspension geometry of a car, it is fairly simple to do a component layout on a scale drawing to achieve this design. The accompanying drawings (starting on the next page) are numbered to correspond to the following steps:

1) Do a drawing in at least 1/4 scale. That means the true dimensions from the race car are divided by

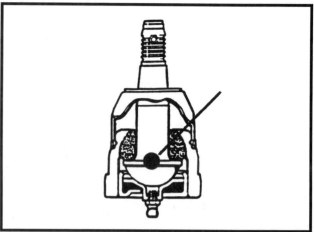

When taking measurements of ball joint pivot points, finding the correct pivot point location is critical. The correct location is the CENTER of the ball of the ball joint. Hold a ball joint in your hand and pivot the stud around so you get a good feel for where the center of the ball is located.

4 when locating them on the drawing. Be very careful when working in this scale because a very small error can make a significant difference.

2) Draw the ground line, vehicle center line, and the center of the left front and right front tire contact patches. Determine where the outer lower control arm ball joint centers are located by bolting the upper and lower ball joints to the spindle and bolting the spindle assembly to the tire and wheel to be used. Mark these ball joint centers on the drawing.

3) Determine the desired roll center location, such as 2 inches above ground and 3 inches to the right of the vehicle centerline. Mark that point of the drawing.

4) We want the instant center of the left front and right front control arms located 2 inches inside of the opposite lower ball joint, so draw a vertical line located 2 inches to the inside of each lower ball joint. We'll call this the instant center vertical plane.

5) Determine the location of the pivot center of each upper ball joint. With the upper and lower ball joints bolted to the spindle, measure from the lower ball joint pivot center to the upper ball joint pivot center. With the spindle and ball joints we used, that measurement was 10.5 inches.

6) Before the location of the upper ball joint centers can be marked on the drawing, the steering axis angle has to be determined. This is the kingpin inclination of the spindle plus the amount of static camber that will be used. The spindle we are using

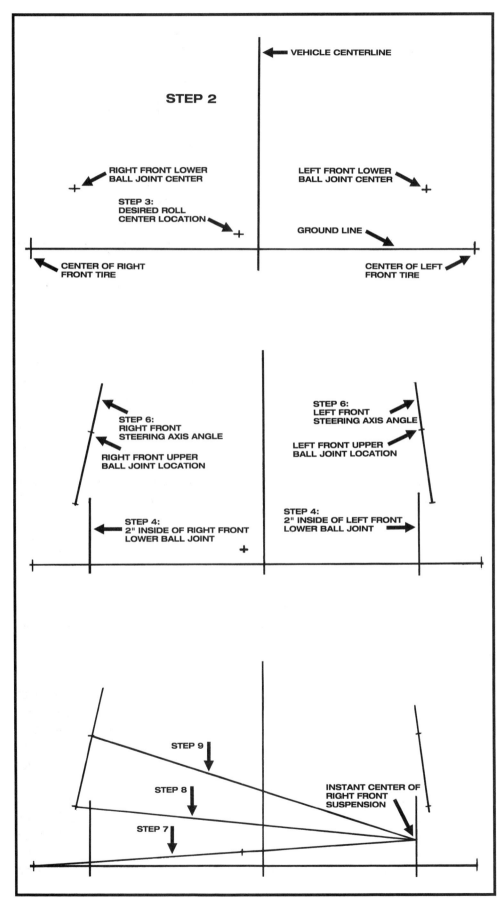

STEP 2

VEHICLE CENTERLINE

RIGHT FRONT LOWER
BALL JOINT CENTER

LEFT FRONT LOWER
BALL JOINT CENTER

STEP 3:
DESIRED ROLL
CENTER LOCATION

GROUND LINE

CENTER OF RIGHT
FRONT TIRE

CENTER OF LEFT
FRONT TIRE

STEP 6:
RIGHT FRONT
STEERING AXIS ANGLE

STEP 6:
LEFT FRONT
STEERING AXIS ANGLE

RIGHT FRONT UPPER
BALL JOINT LOCATION

LEFT FRONT UPPER
BALL JOINT LOCATION

STEP 4:
2" INSIDE OF RIGHT FRONT
LOWER BALL JOINT

STEP 4:
2" INSIDE OF LEFT FRONT
LOWER BALL JOINT

STEP 9

STEP 8

STEP 7

INSTANT CENTER OF
RIGHT FRONT
SUSPENSION

has a 10° kingpin angle, and the initial negative camber setting at the right front is 3°. So, an angled vertical line is drawn from the right front lower ball joint center at 13°. At the left front, a 10° spindle is used and the initial static camber will be positive 1.5° (which tilts the top of the spindle away from the vehicle centerline). So 10° minus 1.5° is 8.5°. Draw a vertical line from the left front lower ball joint center at 8.5°.

7) Draw a line from the center of the right front tire contact patch through the roll center to the instant center vertical plane on the left side. The point where this line touches the instant center vertical plane will be the instant center of the right front suspension.

8) Draw a line from the right front instant center to the right front lower ball joint center.

9) Draw a line from the right front instant center to the right front upper ball joint center.

10) Draw a line from the center of the left front tire contact patch through the roll center to the instant center vertical plane on the right front side. The point where this line touches the instant center vertical plane will be the instant center of the left front suspension.

11) Draw a line from the left front instant center to

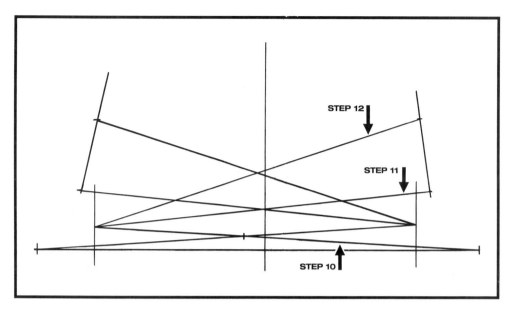

STEP 12

STEP 11

STEP 10

First, you must determine the location of the lower and upper outer pivot points when the suspension is moved 3 inches into bump travel. Draw a horizontal line 3 inches above the lower outer pivot point. Use a compass to swing an arc about the lower inner pivot point, making an arc from the center of the lower ball joint up to meet the 3-inch horizontal line. This intersection is where the lower ball joint center is located when the suspension travels 3 inches in bump.

the left front lower ball joint center.

12) Draw a line from the left front instant center to the left front upper ball joint center.

13) Now that the lines have been drawn from the instant center on each side back to the upper and lower ball joint centers, these lines which converge to the instant center will dictate the planes on which the inner pivot points must be located for the upper and lower control arms on each side. The pivot points must fall on these lines.

14) The only thing left to decide is the length of the upper and lower control arms on each side. The length of these arms will dictate the camber change curve for each side. At the right front, we want to design the arm lengths to achieve 4.25° of negative camber gain in 3 inches of bump travel.

15) The location of the lower control inner arm pivot points will be dictated by the length of the steering rack being used. This is because the inner pivot points of the steering tie rods must be straight in line with the lower control arm inner pivots so that the bump steer is correct. In our example, we are using a rack and pinion that measures 18.25 inches center to center on the rack shaft pivot points. Split that measurement in half (because the rack and pinion is centered on the vehicle centerline), so the lower control arm pivots are 9.125 inches out from the car's centerline on each side.

16) To establish the upper inner pivot point location, you simply work out the required length of the upper control arm. This is done by working backwards from the desired camber gain.

Find where the upper ball joint is located in bump by first drawing a horizontal line 3 inches above the upper outer pivot point. The desired camber gain at the right front is 4.25° at 3 inches of bump travel. Add to this the steering axis angle (13°), which makes 17.25°. Draw an angled vertical line at 17.25° with a protractor, starting the line at the lower ball joint center when elevated at 3 inches. Where this angled vertical line crosses the 3-inch upper horizontal line is the desired location of the upper ball joint at 3 inches of bump travel.

The inner pivot point location for the upper control arm is determined by swinging arcs about different locations on the upper control arm instant center line until the correct angular change is found. The correct angular change will connect the starting upper ball joint pivot location with the intersection of the 17.25° vertical line and the 3-inch horizontal bump travel line. Finding the correct upper control arm length is a matter of trial and error, but start with a commonly used upper arm length. We started with 8.5 inches. Draw a mark in scale on the upper control arm instant center line 8.5 inches inward toward the centerline of the car from the upper ball joint center. Then swing an arc about that mark, from the static upper outer ball joint position up to the 3-inch horizontal line. In our case, that arc intersected perfectly with the 17.25° steering axis line, so an 8.5-inch upper control arm creates the camber gain we were looking for.

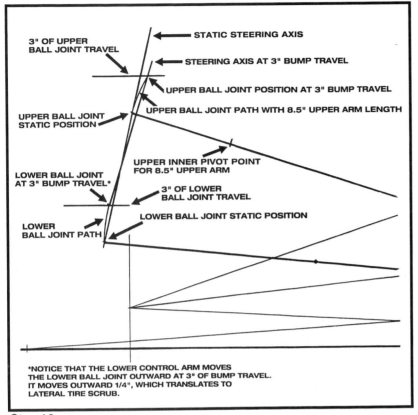

3" OF UPPER
BALL JOINT TRAVEL

STATIC STEERING AXIS

STEERING AXIS AT 3" BUMP TRAVEL

UPPER BALL JOINT POSITION AT 3" BUMP TRAVEL

UPPER BALL JOINT PATH WITH 8.5" UPPER ARM LENGTH

UPPER BALL JOINT
STATIC POSITION

UPPER INNER PIVOT POINT
FOR 8.5" UPPER ARM

LOWER BALL JOINT
AT 3" BUMP TRAVEL*

3" OF LOWER
BALL JOINT TRAVEL

LOWER BALL JOINT STATIC POSITION

LOWER
BALL JOINT PATH

*NOTICE THAT THE LOWER CONTROL ARM MOVES
THE LOWER BALL JOINT OUTWARD AT 3" OF BUMP TRAVEL.
IT MOVES OUTWARD 1/4", WHICH TRANSLATES TO
LATERAL TIRE SCRUB.

Step 16

Lower Control Arm Design

A problem seen sometimes on cars during heavy braking while entering a turn is a high speed chatter at the right front. This is caused by a torsional wrap-up of the linkages that hold the spindle. The high speed chatter is a repeated torquing and un-wrapping of the linkages in high speed cycling. Front brake rotor runout can add to this problem.

The problem is that the control arm and the strut rod are too light to properly anchor the spindle. There is a tremendous torque loading of the lower control arm and its supporting strut rod during heavy braking.

The answer is a proper design of the lower control arm and strut rod, properly sized material to prevent flexing, and good solid brackets that will not flex.

Spindle Choice

The spindle is one part of the overall front suspension geometry design. The pivot points on the spindles (which are the upper and lower ball joints) help to define the roll center of the front suspension. That means a taller or a shorter spindle height will create

a higher or lower roll center. Once a particular spindle is chosen, the balance of the front suspension geometry (upper and lower inner mounting heights and A-arm lengths) has to be designed around it to create the desired roll center height, camber change curve and scrub radius. The tire width and diameter and wheel backspacing will also have an effect on these design elements.

The other elements of spindle design criteria include steering arm length and location, and kingpin axis.

The steering arm length and location determine Ackerman steer, steering quickness and bump steer.

The steering quickness and response is a product of the steering arm length. A shorter steering arm produces faster steering response because the tie rod has to travel a shorter distance to steer the spindle a specified amount. However, if the steering arm gets too short, it prevents the driver from having a smooth controllable feel of the steering. Most aftermarket spindles used for paved track cars have a steering arm length of 4.75 to 5.25 inches. Many times the left front steering arm will be approximately 1/8-inch shorter (or adjustable in length) in order to accommodate Ackerman steering adjustments (more on this principle later).

The steering arm's vertical placement on the spindle affects bump steer. Bump steer is adjusted by

If you look closely, the right front tire is a blur, caused by a high speed chatter under heavy braking

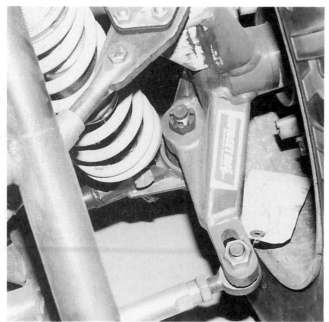

The Coleman aluminum spindle is a modular bolt-together design. It weighs 2.5 pounds less than a standard steel fabricated wide-5 spindle, which removes unsprung weight from the chassis.

Sweet Manufacturing spindles include this Ackerman-adjustable slot on the left front spindle steering arm which allows the tie rod to be set forward or back to adjust the amount of Ackerman.

placing shims between the bottom of the steering arm and the tie rod end. A properly designed steering arm will require a minimum amount of shimming. (Bump steer is covered in-depth later in this chapter.)

Ackerman steer (which is also covered more in-depth later in this chapter) is determined by the mounting angle of the steering arm when compared with the lower ball joint (in top view). If the lower ball joint center and the tie rod mounting center are straight in line with each other and parallel to the centerline of the car, no Ackerman steer is present in the spindle. If the end of the steering arm is moved outward toward the wheel on a front steer car, Ackerman steer is introduced into the spindle design.

The final choice in spindle design is the material. The ultimate choice is a trade-off between cost, weight, reliability and repairability. The most commonly used spindle in paved track racing is a fabricated steel spindle. They are reliable and durable, and because they are fabricated, the design can deliver any type of geometry desired.

The other choice is an aluminum spindle. The major advantage of the aluminum spindle is that it weighs 2.5 pounds less than a standard steel fabricated wide-5 spindle, which removes unsprung weight from the chassis. The Coleman aluminum

spindle is a modular bolt-together design. Each individual part of the spindle is replaceable, which makes this spindle more repairable than a fabricated steel spindle. For the extra cost, aluminum spindles are a good investment in removing unsprung weight.

Kingpin Inclination / Steering Axis Inclination

These two terms are nearly interchangeable, but not quite. Kingpin inclination is the angle from true vertical of a line drawn through the center of the upper and lower ball joint holes of a spindle. Steering axis inclination is the "installed kingpin inclination angle." In other words, it is the operating angle of the spindle after it is installed in the car and camber has been added to it.

On paved track stock cars, the kingpin inclination of the spindles used is generally 10°. This angle is a compromise between the amount needed to reduce the scrub radius to a manageable amount, and the amount needed to minimize the weight jacking effect caused by positive caster during steering. To minimize the scrub radius, more kingpin inclination is better. But to reduce the weight jacking effect, less angle is better.

The steering axis inclination has an effect over how the positive caster angle creates weight loading in the chassis during steering. Positive caster will cause the left front corner to rise up and add weight to that corner and the right rear when the car is steered to the left. When the car is steered to the right, positive caster causes the right front to rise up and add weight

This is a universal upper control arm from PRE. It is infinitely adjustable to fit any spindle height and roll center desired. It slides up and down, and the length can be adjusted. Everything is made of aluminum except the clevises, and every individual piece is replaceable.

to that corner and the left rear. The steering axis angle multiplies this weight jacking effect. The greater the steering axis inclination, the greater the weight loading caused by positive caster.

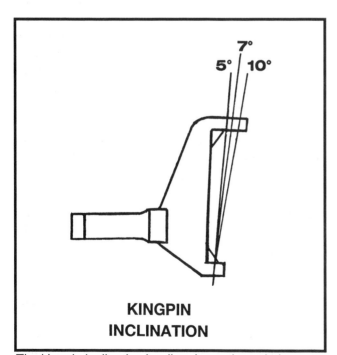

KINGPIN INCLINATION

The kingpin inclination is a line drawn through the upper and lower ball joint centers about which the spindle rotates as it is steered.

Flat-mounted upper control arm mounting brackets offer several advantages. First, these type of brackets replace upright brackets. The upright brackets have mounting holes at fixed positions. The suspension has to work around this. On the other hand, the flat slotted brackets allow a wide range of adjustment. They also offer an easier means of adjustment. Adjustments for camber and caster can be quickly and easily made by moving bolts in a slot. This eliminates the need to add or subtract spacers. The slotted adjustment brackets allow you to get the exact caster and camber numbers you want, not just a compromise close to what you need.

The caster weight jacking effect can be used to help adjust the chassis. If a car is naturally tight, you can get the car to loosen up at turn entry by adding more positive caster at the left front. This creates more weight jacking from the left front to the right rear, and takes cross weight out of the chassis, which loosens it up.

When the steering axis line is projected to the ground, and a measurement is taken from that point to the centerline of the adjacent tire, the amount of scrub radius is determined.

Scrub Radius

The scrub radius is the distance from the upper and lower ball joint projection line (or steering axis line), as it meets the ground, to the center of the adjoining tire contact patch. The amount of wheel backspacing and the steering axis inclination angle of the spindle both have a bearing on the width of the scrub radius. More wheel backspacing, and a larger steering axis angle, narrows the scrub radius.

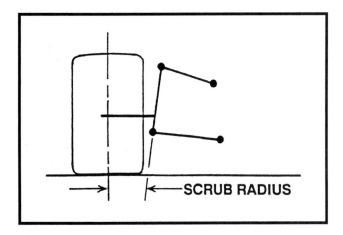

←——SCRUB RADIUS

The scrub radius is the tire's turning radius about the steering axis on the front wheels. As the backspacing of the wheel is increased or decreased, the scrub radius is decreased or increased.

There is a positive aspect and a negative aspect about scrub radius. The positive is that it helps to provide steering feel and feedback to the driver. If there was no scrub radius (which is called center point steering), most drivers would complain that the car reacted too quickly and the steering was darty, and there was very little steering effort and feedback to the driver. The negative aspect is that too much scrub radius causes a lateral scrubbing of the tire, which heats up the tire and increases tire wear.

For short track cars, the amount of scrub radius used is a compromise between the ultimate track width of the vehicle and the amount of tire scrub generated by the scrub radius. A wider track width (which is desirable for less lateral weight transfer) is created by using less wheel backspacing, but that moves the wheel outward and creates a larger scrub radius. Decreasing the scrub radius with more backspacing narrows up the scrub radius, but that also narrows the track width by moving the tire inward toward the centerline of the car.

With a scrub radius, when you turn left, it lengthens the right side wheelbase, which tends to loosen the car. When you countersteer, it shortens the right side wheelbase and it puts understeer in the car. This is a very stabilizing effect to the control of the car and in feedback to the driver.

When no scrub radius is present, it creates center point steering, which has zero scrub on the front tires. This happens when the steering axis inclination intersects the center of the tire contact patch. With this situation the driver has no seat-of-the-pants steering feel or feedback at all. The driver will complain about the front end moving around and he can't feel what the car is doing. Many racers who have never done a scale drawing of their chassis have this problem, and have made other changes (spring rates, excessive positive caster, etc.) to try to correct the situation.

Most cars, with 10 to 11 inches of tire tread face on the ground and using a 10-degree inclination spindle, will have a 4.5 to 5-inch scrub radius at the right front, depending on wheel backspacing. The left front will be about 1 to 1.5 inches wider because of the positive camber used on that wheel.

The scrub radius is extremely important to the driver for good feedback on how the car is handling. This feedback tells him how heavily the tires are loaded and when they are at the edge of their traction ability. The scrub radius should be between 3.625 inches and 6 inches. The ideal right front scrub radius is 5 inches. If it falls below 3.625 inches, it does not produce enough driver feedback. Past 6 inches, it adds too much tire scrub about the steering axis.

Using less than a 10° inclination angle spindle will increase the scrub radius. Switching from a 10° to a 5° spindle at the right front will increase the scrub radius by nearly an inch.

Spindle Overall Height & Design

The overall height of the spindle upright is another design consideration for front end geometry. A shorter spindle upright height (upper ball joint mount to spindle pin center line is shorter), with all other pivot points and ground clearance remaining the same, will lower the roll center height. This shorter spindle will produce a lower roll center height because the geometry creates a longer instant center. The shorter the instant center to the wheel it is drawn from, the higher the roll center will be.

Because paved track cars require a lower front roll center, they use a shorter spindle than dirt track cars. The most common front spindle height used for paved track cars is 7.75 inches.

The location of the ball joints in relationship to the spindle's pin location makes a difference on lateral displacement of the tire during bump and rebound travel. Lateral displacement is bad for a tire because the tire slides outward against the track surface, resulting in excessive tire heat and wear.

This front suspension design is unique because it allows the racer to use a coil-over or a conventional spring, depending on the rules at the track being run. (Right) To use a conventional spring, the coil-over is removed and a conventional weight jacker is installed, anchored to the upper control arm inner bolts and the chassis loop above it.

achieved when the top of the steering axis is inclined toward the front of the vehicle in the side view. Positive caster creates straight ahead stability, preventing wander. Positive caster puts a positive feel into the steering for the driver.

Positive caster provides a self-centering effect to the steering (called self alignment torque). The more positive caster, the more tracking feel there is for the driver. However, the more caster used, the harder the driver's steering effort becomes when turning the car.

Negative caster creates the opposite of a self aligning torque. Where the positive caster tries to resist a steering effort and align the wheel straight, negative caster will help the wheel to deviate from a straight path. Negative caster should not be used.

While positive caster provides a more positive steering feel, it also has some negative implications for the chassis as well. When a car with positive caster turns left, the left front corner will rise and the right front corner will dip. The amount of these changes depends on the amount of positive caster used combined with the spindle's steering axis inclination angle. The steering axis inclination multiplies the effect of the positive caster and the associated

corner lift and drop. The greater the steering axis inclination, the more that positive caster will change the corner height of the car as the wheel is steered. This effect is caused by the curved path that the spindle pin follows as it is turned about the steering axis.

As a car with positive caster is steered left and the left front corner rises, the result is the same as jacking weight into that corner. The chassis gains weight at the left front and right rear corners, and loses weight at the right front and left rear. This effect takes some cross weight out of the chassis. The more positive caster used at the left front and the greater the steering axis angle, the greater the loss of cross weight in the chassis as it turns left.

Another situation present in some chassis is caster gain or loss as the car corners. This depends on how the upper and lower A-arms are mounted and angled. If the rear mount of a lower A-arm is mounted lower than its front mount (in side view), the bump and rebound path of the lower ball joint will be on a radius instead of a straight up and down movement. This can cause caster gain or loss as the wheel moves through bump and rebound. This will cause

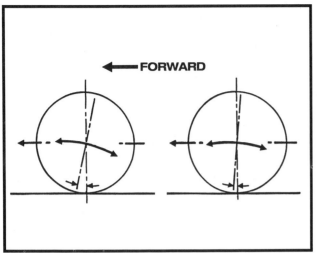

When a car with positive caster turns left, the left front corner will rise. The amount depends on the amount of positive caster used combined with the spindle's steering axis inclination. The steering axis inclination multiplies the effect.

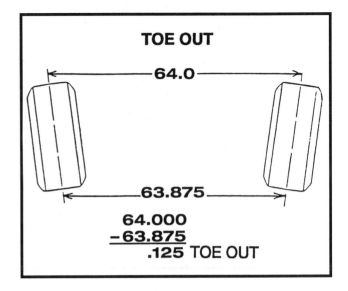

a change in steering feel for the driver, as well as jack weight in and out of the chassis.

On a fabricated front clip, zero caster gain or loss is easy to achieve. The inner mounting points for the upper and lower A-arms should be mounted parallel to each other and parallel to the ground, looking at the car in side view. And, looking at the car from the top view, the inner mounting points should be parallel to each other and parallel to the vehicle centerline.

Caster Split

Caster split is the difference between the amount of left front and right front caster. When caster angles are not equal, the steering will tend to pull toward the side with the least amount of positive caster or the side with more negative caster. On stock cars on an oval track, a greater amount of right front positive caster than left will enable the driver to steer the car into a corner with less effort. The greater the amount of caster between the left front and the right front, the greater the caster split.

Caster split can be anywhere from 1 to 4 degrees. Short tight tracks can use more caster split to help the car make a tight arc with minimal effort. However, more caster split makes it much more difficult to turn the car back to the right. This makes it hard to maneuver in traffic, to counter steer, and to drive out of trouble. Power steering helps minimize the problems associated with more positive caster and

more caster split. The bottom line on stagger: make it comfortable for the driver. Most drivers are comfortable with no more than a 2 to 3-degree split.

The amount of positive caster used at the left front depends upon the length of the race track. A shorter, tighter track requires less positive caster at the left front, which creates more caster split. This helps pull the car into the turn.

On a faster half-mile or larger track with wider turns, a little more left front positive caster yields less caster split. This gives the car more stability on the straightaways. But more left front positive caster also will jack more weight from the left front to the right rear at turn entry, which will loosen up the chassis. In this case, the chassis should be set up with a slight initial understeer.

The proper range of caster adjustment is discussed in the Chassis Adjustment chapter.

Toe-Out

Toe-out is the difference in distance between the front and rear measurements of the tires on the same axle in the center of the tread surface, at spindle height, where the front measurement is greater than the rear. Toe-in is the opposite.

Toe-out is a static alignment made to minimize tire scrub and rolling resistance which develop through the tires when a car is cornering. When a car is in a turn, both front wheels are turning about a common center, but because the left front is closer to that center, it is turning on a sharper radius. Thus with the two front wheels of the car turning on separate radii, the two tires are actually pointing in different directions with the left wheel making the sharper turn. The purpose here is to introduce more initial toe-out so each front wheel follows its own radius in a perfect path through the turn. When either front wheel is at an angle against its turning radius, it is requiring extra horsepower to scrub that tire through the turn. The amount of toe-out used should be kept to a minimum in order to minimize tire scrub down the straightaways.

The way that the toe-out measurement is taken is critical, because proper toe-out is measured in small increments. 1/16 of an inch can make a big difference. Because of that, a good measuring technique is important.

Using a tape measure spread between the two front wheels is not a got idea. The play or flex in a tape measure over a 64-inch length can create more than a 1/8-inch error in the measurement. Use a tape only in a pinch. The best way to measure toe is with a trammel bar. This procedure, and the proper range of toe-out adjustment, is described in detail in the Chassis Set-Up chapter.

Checking Toe-Out As A Preventive Measure

If your race car is involved in any type of wreck, altercation or wall bumping incident, checking the toe-out is a very quick way of determining if anything in the front end is bent. But, the most important thing is knowing what the toe-out was when your car went out on the track. Always keep track of it. If you make serious front wheel contact with something on the track, and then measure the toe and it has changed, something in the front end is bent.

Toe-out measuring is important. Don't overlook it.

Anti-Dive

Anti-dive is the use of mechanical means to keep the nose of the car from diving under braking loads. It is caused by dissimilar mounting angles of the upper and lower A-arms (in the side view). Anti-dive is used to resist forward longitudinal forces by jacking up the front of the car when the forces are applied. Under 100 percent anti-dive, no nose dive will occur under heavy braking.

However, these factors produce undesirable side effects in the chassis. The mechanical bind caused by anti-dive will start to lock up the front A-arm pivot

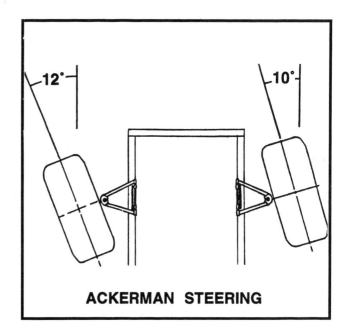

ACKERMAN STEERING

points under braking and forward weight transfer. This will add extra spring rate to the front end, and cause front wheel chatter under braking.

The large amount of caster change caused by anti-dive is an unsettling element to the driver, because the caster is totally dependent on the amount of bump travel the wheel is experiencing.

The best bet is to build the car with zero anti-dive. Follow the same guidelines for the inner upper and lower A-arm mounting as discussed previously for avoiding caster gain and loss.

Ackerman Steering

Ackerman steering geometry is created when the inside front wheel is steered in a sharper arc than the outside front in order to eliminate tire scrub at the inside wheel during cornering.

The Ackerman steering principle can be somewhat controversial with some people because decades ago some noted British engineers designed some race car suspensions with zero Ackerman or anti-Ackerman steering. While it may have worked then, technology has changed. The predominant change is in tire technology. Today's tires have a higher coefficient of friction, tire widths are wider, and the tires operate at much less of a slip angle. This coupled with the fact that a proper short track setup keeps the left front tire loaded vertically means that Ackerman steering must be used to optimize the steering angle of the left front tire.

You can see Ackerman steering at work here, with the left front steering at a greater angle than the right front.

The left front is a very important tire contact patch. And thus, the direction it steers without tire scrub is very important. It must steer at a sharper arc than the right front because it is travelling on a shorter radius. The left front simply steers more and helps point and stabilize the car. Increased Ackerman decreases — or eliminates — cornering understeer at turn entry and through the middle of the corner. It responds this way in both high speed and low speed cornering.

Ackerman steering actually is dynamic front toe-out. It only creates toe-out as the front wheels are steered. Large amounts of toe-out are required to steer a car into a turn properly, but large amounts of static toe-out are not beneficial to the car. It creates excessive drag on the straightaways and causes darty steering response. However, Ackerman steering will only create toe-out when the car is steered. Ackerman is gained directly in proportion to how much the steering wheel is turned.

It is important to have Ackerman steering (toe-out gain) when the car is steering both to the left and to the right. This is so that the front suspension will not create toe-in when the car is counter-steered to the right. When a car is counter-steered and the wheels are turned to the right, Ackerman will steer the right front tire more outward. This will help steer the nose of a car out of a spin.

How Much Ackerman Steering?

The amount of Ackerman steering a car has is measured in the difference in steering angle between the right front and the left front wheels. Because on oval tracks the right front is the controlling tire, the amount of Ackerman steering is quoted as the amount of steering angle gain at the left front over the right front.

In general, the left front has to turn about 15 percent more than the right front. This means that if the right front tire turns 10 degrees, the left front is going to have to turn at least 11.5 degrees to minimize tire scrub. Therefore, a paved track car requires 3 to 4 degrees of toe-out gain at the left front in 18 degrees of right front steering. In other words, if you turn the right front wheel 18 degrees to the left, the left front should show 21 to 22 degrees of steering angle. The tighter the turn radius, the more the left front has to turn. The wider the turns – like on a wide sweeping 1/2-mile track – the less Ackerman gain required at the left front.

Ackerman steer is created on a front steer car by having the steering arms angled outward toward the brake rotors. But, there are other ways of gaining Ackerman also. One way is through carefully controlled bump steer. For short track cars, bump steer should be 0 (no steer whatsoever through bump and rebound travel) at the left front and should be 0.035-inch to .040-inch out per inch of upward spindle travel at the right front (on a flat or banked track).The reason for this is that the right front of the car is the controlling side for steering. The deeper you drive into a turn, the more bump (upward) travel there is at the right front. And the closer you get to the apex of the turn, the more the car is going to push.

The trick is to get the car to turn at the apex without pushing so you can get back on the throttle immediately. The key to getting the car to turn is more toe-out at this point. The more toe-out you have here, the more the car will cut down without pushing. As the right front bumps out, what is really happening is the driver can steer the car to the left more as he keeps the right front in line with the radius of the turn. So as the car rolls over it is trying to bump out at the right front, the driver is steering more to the left, and this is creating more toe-out at the left front. The toe-out is really showing at the left front even though it is bumping the right front out. This is a controlled bump out because the more the right front of the car is compressed, the more it toes out. This is much better than static toe-out, which can be detrimental on the straightaways.

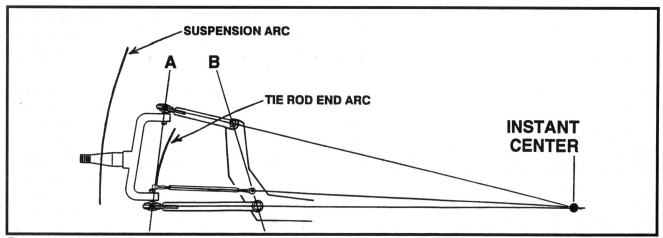

Bump steer is zero when the suspension arc and the tie rod end arc are parallel to each other. To accomplish this, the tie rod length must fall between the confines of planes A and B, and the tie rod end's centerline must intersect the instant center.

Using Ackerman steer in the chassis not only is helpful for turning the car into the turn, but also for countersteering when a car is loose. When you countersteer and have Ackerman present in the steering geometry, the right front will toe out and it will point the right front outward and stabilize the car. However, if there is no Ackerman or the front wheels toe in when countersteered to the right, the front end of the car will be pinned and you will spin out.

You do not want to gain Ackerman by changing the length of one steering arm (such as by shortening the left front). This will open up a whole world of new problems. One of these problems is that with a shorter left front steering arm, the car will gain toe out, as you want, turning left, but it will gain **toe in** when counter-steering back to the right.

Checking Your Ackerman

To find out how much Ackerman you have built into the steering geometry, you need to have steering plates that measure the number of degrees each wheel turns.

Steer the car to the left. At 5 degrees, 10 degrees, 15 degrees and 18 degrees of left steer of the right front tire, mark down on a chart the number of degrees the left front is steered. Make a comparison between the steering angle of each wheel at each increment. The difference will be the amount of Ackerman present. Be sure to check the wheels steered to the right as well so you can determine if the right front toes in or out during countersteer.

Bump Steer

Bump steer is the change in steering angle of the front wheels caused by the front suspension moving up or down through its travel. Bump steer causes the introduction of toe (either in or out) into the front wheels when the suspension goes into bump or rebound. Bump steer occurs when the tie rod end follows a path different from the path the wheel is following. If you visualize the arc created by the front wheel as it moves up and down through its travel, along with the arc created by the tie rod end, you will see that both arcs must have the same instant center or the tie rod end will move in or out in relation to the wheel, causing a change in toe. If the arc created by the tie rod end and the wheel are parallel, no toe change will occur. Bump steer results from the steering tie rod moving in a path which is dissimilar to the path of the connected wheel.

For short track cars, bump steer should be 0 at the left front (no steer whatsoever through bump and rebound), and 0.035-inch to 0.040-inch out per inch of upward spindle travel at the right front, on both flat and banked tracks. The more toe-out bump steer there is at the right front, the less static toe-out the car requires. This is because the bump steer helps create the toe-out as the body rolls during cornering.

Setting Bump Steer

To check and set the bump steer, first set the caster and camber at the correct static alignment. Set the toe-out to zero. Clamp the steering wheel tight with the front wheels in the straight ahead position.

This bump steer gauge from REB-CO uses only one dial indicator, so results read directly without the need for subtraction.

(Check the centering of the steering by making sure that the inner tie rod pivot points are centered in relation to the inner pivot points of the lower A-arms.) Then place the car on jack stands or wood blocks to keep the chassis blocked at its correct racing ground clearance at all four corners.

Disconnect the springs and shocks. Remove the wheels and bolt a flat plate to the hub. This flat plate should have a hole cut in its center through which the hub protrudes so the wheel studs can bolt through corresponding holes drilled in the flat plate. The plate should have lines scribed or drawn on it at 1/2-inch increments so the amount of travel through bump or rebound of the spindle can be accurately measured.

Two dial indicators that read in thousandths of an inch are attached to a stand which will hold them against the flat plate as the suspension is moved through bump and rebound. Place a jack under the lower A-arm and bring it to normal ride height. Make sure the flat plate is level. Use at least one-inch travel dial indicators. When the dial indicators are placed against the plate at the 0 travel position (normal ride height), compress the indicator needles against the

Bump Steer Patterns
For Front Steer Suspensions

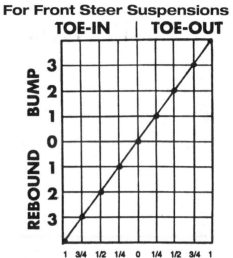

The wheel is toeing out on bump and in on rebound. The cause is the outer tie rod end is too low or the inner is too high.

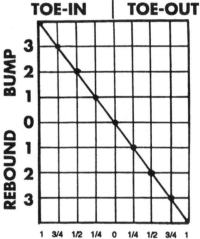

Toeing-in on bump and out on rebound indicates the outer tie rod end is too high or the inner tie rod end is too low.

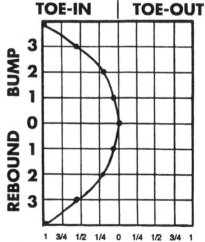

A convex curve like this indicates the tie rod end heights are compatible but the tie rod is too short, causing the wheel to toe-in on both bump and rebound.

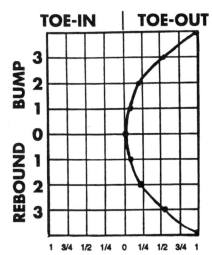

The tie rod here is too long, causing toe-out on both bump and rebound.

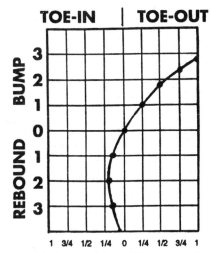

This is a combination problem. The outer tie rod end is too low and the tie rod is too long.

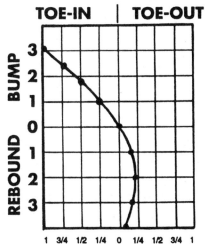

Another combination problem. The outer tie rod end is too high and the tie rod is too short.

plate slightly and then set both dial indicators to zero. Make sure that both indicators are set equal distance from the plate.

At the right front, use the jack to move the suspension through three inches of bump (upward) travel, and one inch of rebound (down) travel. At the left front, move the spindle through two inches of bump travel and three inches of rebound. Starting at the right front, move the suspension 1/2-inch at a time and make a chart of the bump steer. The difference between the readings on the dial indicators is the measurement of bump steer. Be sure to note if the wheel is toeing in or out. As you measure the toe change at increments of bump and rebound, plot the results on a chart like we have illustrated here. The patterns on the chart will give you an indication of the problem causing the bump steer. All of the charts shown here are for front steer cars.

On fabricated front clip race cars, the majority of bump steer problems can be solved by the proper placement of the steering rack. Mount the rack so that the inner tie rod ends are centered vertically and horizontally with the lower inner A-arm mounting points. Bump steer can then be corrected by shimming the outer tie rod ends up or down where they mount to the spindle steering arms. If necessary, the rack can be shimmed up or down to help perfect the bump steer.

Making suspension changes to correct the bump steer can be a time consuming process. But it is a very important suspension adjustment. Make sure you take the time to do it. And, any time your race car has suffered any damage to the front suspension, steering or frame rails, be sure to do a bump steer check again.

A much easier and quicker way to check the bump steer on a car is with the ***Racing Chassis Analysis*** computer program available from Steve Smith Autosports. Once specified measurements are taken on the chassis and entered into the computer program, both a bump steer chart and bump steer graph are generated. After reviewing this data, the racer can go back into the program and change measurements (such as relocating inner or outer tie rod connecting points, etc.) to fine tune the bump steer curve until the results show the desired curve. Then, armed with these ideal mounting points, he can go back to the shop and make the required changes in his suspen-

			Supension Bump				
Left Wheel Travel	Left Camber	Left Camber Change	Left Bump Steer	Right Wheel Travel	Right Camber	Right Camber Change	Right Bump Steer
4.000	-2.188	-0.921	0.052	4.000	-9.373	-1.463	0.070
3.500	-1.513	-0.856	0.040	3.500	-8.492	-1.415	0.057
3.000	-0.874	-0.790	0.028	3.000	-7.611	-1.365	0.045
2.500	-0.342	-0.730	0.019	2.500	-6.840	-1.320	0.035
2.000	0.155	-0.668	0.012	2.000	-6.071	-1.273	0.025
1.500	0.584	-0.607	0.006	1.500	-5.354	-1.227	0.017
1.000	0.950	-0.548	0.002	1.000	-4.689	-1.183	0.010
0.500	1.255	-0.490	0.000	0.500	-4.071	-1.139	0.004
0.000	1.500	0.000	0.000	0.000	-3.500	0.000	0.000
-0.500	1.697	0.373	0.002	-0.500	-2.942	1.053	-0.003
-1.000	1.824	0.317	0.006	-1.000	-2.462	1.012	-0.005
-1.500	1.895	0.261	0.012	-1.500	-2.023	0.972	-0.005
-2.000	1.908	0.204	0.020	-2.000	-1.626	0.933	-0.004
-2.500	1.859	0.143	0.030	-2.500	-1.249	0.891	-0.002
-3.000	1.742	0.080	0.043	-3.000	-0.936	0.852	0.001
-3.500	1.553	0.015	0.058	-3.500	-0.648	0.811	0.006
-4.000	1.288	-0.053	0.075	-4.000	-0.407	0.769	0.013

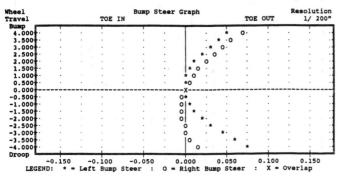

```
Wheel                    Bump Steer Graph                 Resolution
Travel          TOE IN                        TOE OUT       1/ 200"
Bump
 4.000|    .    .    .    .    .    .  *   o·   .    .    .
 3.500|    .    .    .    .    .    .* ·o   .    .    .    .
 3.000|    .    .    .    .    .  · .* o·   .    .    .    .
 2.500|    .    .    .    .    .    .* o   .    .    .    .
 2.000|    .    .    .    .    . *  o   .    .    .    .    .
 1.500|    .    .    .    .    .*o   .    .    .    .    .
 1.000|    .    .    .    .    * o   .    .    .    .    .
 0.500|    .    .    .    .   *o   .    .    .    .    .
 0.000|----------------------X-------------------------
-0.500|    .    .    .    .  o*   .    .    .    .    .
-1.000|    .    .    .    .  o| *  .    .    .    .    .
-1.500|    .    .    .    .  o| *   .    .    .    .    .
-2.000|    .    .    .    .  o|  *  .    .    .    .    .
-2.500|    .    .    .    .   o  *   .    .    .    .    .
-3.000|    .    .    .    .   o   *·  .    .    .    .
-3.500|    .    .    .    .  |o     *   .    .    .    .
-4.000|    .    .    .    .  o      *    .    .    .    .
Droop
         -0.150   -0.100   -0.050   0.000   0.050   0.100   0.150
   LEGEND:  * = Left Bump Steer  :  O = Right Bump Steer  :  X = Overlap
```

*These bump steer graphs were generated using Steve Smith Autosports' **Racing Chassis Analysis** computer program.*

sion. The bump steer graphs shown were generated using the ***Racing Chassis Analysis*** program.

Spring Rate Vs. Wheel Rate

The spring rate of a spring is the comparative rating of one spring against another in terms of its resistance to a load placed on it. This measurement is expressed in terms of pounds per inch. For example, if 200 pounds were placed on a spring and it compressed one inch, the spring rate is 200 pounds per inch (load divided by inches of compression).

The wheel rate of the spring is the effective rate of the suspension spring at the lower ball joint of the A-arm which compresses the spring (on an independent front suspension). The wheel rate is the spring rate corrected by the mechanical advantage, or leverage. This correction factor is the motion ratio of the linkage squared. The motion ratio, or leverage ratio, is the pivot point to spring center distance (A on the drawing) divided by the total effective length of the A-arm (B in the drawing).

Wheel Rate Vs. Wheel Load Rate

Wheel rate, at the front of an independent suspension car, is the measurement in pounds per inch of

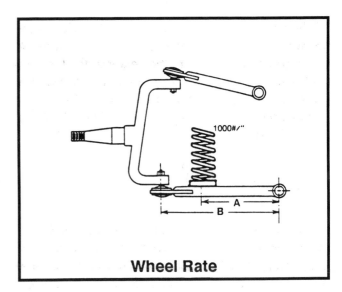

Wheel Rate

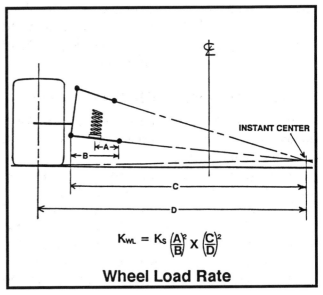

$$K_{WL} = K_S \left(\frac{A}{B}\right)^2 \times \left(\frac{C}{D}\right)^2$$

Wheel Load Rate

effective spring rate at the lower ball joint. The effective spring rate is calculated at this point because the lower ball joint is the outer most connecting point of the lever arm (lower A-arm) which creates the leverage ratio or motion ratio that acts upon the spring rate. The wheel rate (or ball joint rate) has been used for years as a common denominator in comparing different springs with the same leverage ratio, or the same spring with different leverage ratios.

Many racers notice that the ball joint is located a distance away from the center of the tire contact patch, and the distance from the ball joint to the tire center varies among different cars or with differences in wheel backspacing. Thus they are concerned about determining the spring rate effect at the center of the tire and not at the ball joint. This rate, called the wheel load rate, takes into consideration the total lever arm extending from the center of the tire contact patch to the inside pivot point. It measures the total effective spring rate at the center of the tire contact patch, or in other words, the wheel loading rate. It is computed by the following formula:

$$K_{WL} = K_S \left(\frac{A}{B}\right)^2 x \left(\frac{C}{D}\right)^2$$

As you can see, the formula is only a modification of the wheel rate formula. And, if you plug in actual numbers to the wheel load rate formula, you will find that the answer varies **very** slightly from the wheel rate, even with wide differences in wheel offset. And, you can thus see that, for ease of computation and keeping things simple, it is easier and simpler to just use the wheel rate (or ball joint rate).

Wheel Rate With Coil-Overs

The wheel rate with a coil-over spring is determined very similarly to a conventional coil set-up, with one element added.

The motion ratio for a coil-over spring is determined by dividing the coil-over mounting point on the lower A-arm length (A in the drawing) by the total A-arm length (B in the drawing), then squaring this number. Then it is multiplied by the cosine squared of the mounting angle. This is necessary because the mounting angle of the spring decreases the leverage on it.

Use an angle finder to find the number of degrees from vertical that the coil-over is mounted. Then find the cosine of that angle in a Trig Table, square it, and multiply that times the motion ratio squared. A Trig Table can be found in the back of **The Racer's Math Handbook** and **Advanced Race Car Suspension Development,** both published by Steve Smith Autosports. Or, it can be found in math books, or even some hand calculators have that function built into them. The cosine for most common coil-over mounting angles is included here in a chart for you.

As an example of a coil-over wheel rate computation, assume that the lower control arm length is 17 inches, the coil-over is mounted 13 inches out from the control arm inner pivot point, the coil-over is mounted 15 degrees inward from true vertical, and

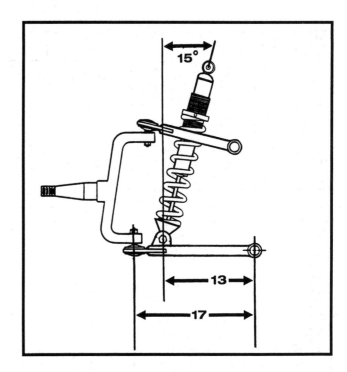

Angle	Cosine
5	0.996
6	0.995
7	0.993
8	0.990
9	0.988
10	0.985
11	0.982
12	0.978
13	0.974
14	0.970
15	0.966
16	0.961
17	0.956
18	0.951
19	0.946
20	0.940

a 500 #/" coil-over springs is being used. To determine the wheel rate of the spring, first you have to find the correction factor:

$$\left(\frac{A}{B}\right)^2 x (\cos)^2$$

$$\left(\frac{13}{17}\right)^2 x (.966)^2$$

$$.765^2 x .966^2$$

$$= .546$$

Multiply this correction factor by the spring rate of 500 #/", and the wheel rate is 273#/" (500 x .546 = 273).

Front Tire Stagger

There is no such thing as front stagger, because the front suspension is independent — the two front tires are not tied together on a common axle. Each tire is independent of each other and a circumference change at one will not affect the other. The only change that will be apparent is that a larger circumference tire is taller than a smaller one, so there will be a corner height difference, which is the same as weight jacking. Most racers, however, refer to this difference as tire stagger. Generally, a left front tire with a circumference one inch smaller than the right front is used on most set-ups. This front tire size difference affects the amount of braking torque applied to each front wheel going into a turn. As brake torques are applied differently, the left front receives more braking torque and pulls the car into the corner. This is the same effect as caster split. A good combination is to use a little more front tire stagger (minimum 1-inch, maximum 1.5 inches) and use a little less caster stagger (1.5° to 2°). What this does is allow you to achieve the same effect as more caster split without the negative effect of massive steering effort back to the right. In addition, the taller right front tire adds more wedge to the left rear.

Anti-Roll Bars

Anti-roll bars (also called stabilizer bars or sway bars) are used to help control body roll and to balance out the front-to-rear cornering weight transfer.

An anti-roll bar functions like a transversely mounted torsion spring, adding spring rate stiffness to the chassis during cornering. As the chassis experiences body roll, the outside arm of the bar moves up while the inside arm moves down, creating a twist in the bar. The resistance of the bar to this twisting movement is the bar's spring rate. The spring rate of the anti-roll bar adds spring rate (during cornering) to the end of the car to which it is attached, which would be the same as adding stiffer suspension springs. The added stiffness means that that end of the car will carry more of the load transfer resulting from cornering.

The anti-roll bar provides a means of controlling body roll separately without interfering with spring rates and vertical wheel control.

An anti-roll bar separates roll control from ride control. If only the springs were available at the front end to control bump movement, downforce loading and body roll, the spring rates would have to be too stiff for optimum handling and optimum wheel control over bumps. Therefore it becomes necessary to provide a means of controlling body roll separately without interfering with the spring rates and thus the vertical control of the wheels. The anti-roll bar is added to the suspension for roll control. Of the total front spring rate (both front springs plus the anti-roll bar), 30 to 50 percent of that total should be from the anti-roll bar. A flat race track would require the most anti-roll bar rate, while a higher banked race track would require the least anti-roll bar rate percentage. That is because the banking angle of the track produces more downforce on the chassis than body roll, and this extra downforce must be controlled with spring rates.

The rate in pounds per inch of arm deflection of a solid anti-roll bar can be found using the formula:

$$\frac{1{,}125{,}000 \times D^4}{L \times A^2}$$

where
D = bar outside diameter
L = effective bar length
A = effective arm length

To compute the rate of a hollow or tubular anti-roll bar, use the formula:

$$\frac{1{,}125{,}000 \times (D^4 - d^4)}{L \times A^2}$$

where
D = bar outside diameter
d = bar inside diameter

There are three variables that affect the spring rate, or torsional stiffness rate in inches of arm deflection, of an anti-roll bar: the diameter of the anti-roll bar along its effective length, the effective length of the anti-roll bar attachment arm, and the effective length of the bar.

Anti-roll bars are very sensitive to changes in diameter. The formula for calculating the rate of an anti-roll bar shows that the diameter is taken to the fourth power (multiplied by itself four times). So a very small change in bar diameter creates a very large change in spring rate. Look at the adjoining table of common anti-roll bar sizes and rates. With a 36-inch effective length solid bar and 12-inch arms, increasing the diameter from 1.0" to 1.25" increases the rate from 217 lbs./inch to 530 lbs./per inch.

The effective bar arm length is the distance from the center of the bar diameter to the center of the arm's attachment point to the lower control arm, measured at a 90° angle to the anti-roll bar (as seen in the side view).

The torsional stiffness of the anti-roll bar gets stiffer as the arm length is shortened. Most generally, stock cars are designed for a 12-inch arm length. There are some applications that have been designed for an arm length of 8 inches.

The arm length can be a very effective bar rate tuning device when the bar is attached to the chassis with a sliding clamp bracket. Howe Racing Enterprises makes a sliding mount kit of this type which mounts a universal-type mandrel-bent bar. The sliding anti-roll bar clamp moves forward or back on the frame rails, effectively shortening or lengthening the bar arms. With a 1.125-inch O.D., .125-inch wall tubular bar, when the mount is set for 13-inch long arms, the rate of the bar is 187 lbs./inch. When the arms are shortened to 8.75 inches, the bar rate is 415 lbs./inch.

The effective length of an anti-roll bar is its active twisting area located between the mounting clamps on either side. For a universal-type mandrel-bent bar, the length is measured just inside of the radius bend to the arms. On a splined bar with separate

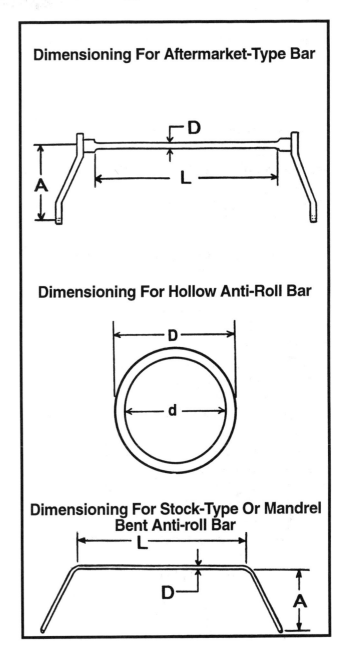

Dimensioning For Aftermarket-Type Bar

Dimensioning For Hollow Anti-Roll Bar

Dimensioning For Stock-Type Or Mandrel Bent Anti-roll Bar

arms, the effective length is the area inside of the splined sections or the turned-down section.

The shorter the effective length of an anti-roll bar, the stiffer its rate is. The proper length of the bar is the outside-to-outside width of the front stub frame rails, with enough extra distance added for the arms to clear the rails. If the effective length of the bar extends too far outside of the bar's mounting brackets, the bar will flex or bend, which creates a bind.

Tubular Anti-Roll Bars

Tubular anti-roll bars are almost universally used on paved track stock cars because they save weight

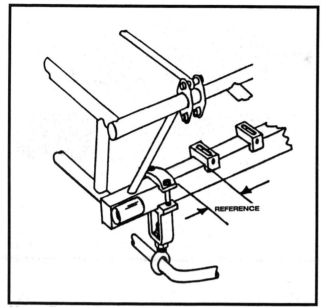

Howe Racing Enterprises makes a sliding mount kit which mounts a universal-type mandrel-bent bar. The sliding anti-roll bar clamp moves forward or back on the frame rails, shortening or lengthening the bar arms and changing the bar spring rate.

Typical Anti-Roll Bar Rates

32" Effective Length, 12" arms, 1-piece mandrel-bent universal bar

O.D.	I.D.	Wall	Spring Rate	Weight
1.0625	.822	.125	200	6.05
1.125	.875	.125	248	6.68
1.25	1.0	.125	352	7.51
1.0	Solid		244	13.35

36" Effective Length, 12" arms, 1-piece mandrel-bent universal bar

O.D.	I.D.	Wall	Spring Rate	Weight
1.0625	.822	.125	178	6.45
1.125	.875	.125	220	7.12
1.25	1.0	.125	313	8.01
1.0	Solid		217	14.24
1.25	Solid		530	22.25

36" Effective Length, 8.75" arms, 1-piece mandrel-bent universal bar

O.D.	I.D.	Wall	Spring Rate	Weight
1.125	.875	.125	415	7.12

36" Effective Length, 13" arms, 1-piece mandrel-bent universal bar

O.D.	I.D.	Wall	Spring Rate	Weight
1.125	.875	.125	187	7.12

– important weight forward of the front wheels. Look at the adjacent chart. A 36-inch long 1.0-inch O.D. solid bar and a 1.25-inch O.D., .125-inch wall tubular bar are almost identical in spring rate. But the solid bar weighs almost double what the tubular bar does.

Anti-Roll Bar Installation

The anti-roll bar has to be mounted square to the lower control arm mounting points in order to work properly. The bar has to be absolutely perpendicular to the centerline of the race car. If the bar is mounted at a slight angle, this will introduce a bind into the bar as it twists, which will add to the torsional stiffness and make it act like a variable rate spring.

To get the most effective spring rate from the bar, the arms must be mounted on the lower control arms as close to the ball joint as possible. The bar arm mounting position is subject to the same type of motion ratio leverage that affect springs. The effective wheel rate of the anti-roll bar is found by multiplying the motion ratio squared by the anti-roll bar spring rate. The motion ratio is the arm mounting position length on the control arm (A in the drawing) divided by the lower control arm length (B in the drawing). Then this motion ratio is squared.

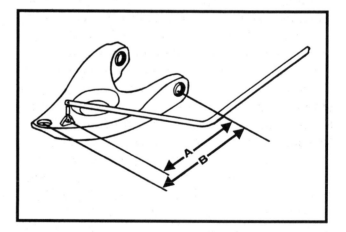

Once the installation of an anti-roll bar is completed, check the bar mounting and linkages through a full range of suspension bump and rebound travel. Look for any type of binding of the mounting linkage, and for proper clearances of all components.

Use 5/8-inch rod end bearings (one male, one female, per side) to attach the anti-roll bar arms to the lower control arms.

Chassis Tuning With The Anti-Roll Bar

Anti-roll bars are a great aid in tuning the roll couple distribution of the race car. Roll couple is the distribution of cornering weight transfer between the front suspension and the rear suspension of the car. The end of the car with the highest roll stiffness will receive the greatest amount of weight transfer during body roll. Because an anti-roll bar adds spring stiffness to the overall combined front end suspension stiffness, the rate of the bar can be fine-tuned to add or subtract spring stiffness in the front during body roll. Adding spring stiffness would make the car tend toward understeer. Subtracting stiffness would make the car tend toward oversteer. Adjusting the spring rate of the anti-roll bar is a very quick and effective method of fine-tuning the front roll couple stiffness.

See the **Track Tuning & Adjustment** chapter for more information on tuning the roll couple distribution.

Rack And Pinion Steering

The major question with rack and pinion steering is how to choose the proper unit for your application. As far as brand choice, there are many companies that make rack and pinion units — Appleton, Coleman, Sweet, and Woodward are among the major ones. Each company produces an excellent product. The final decision, then, should be based on who is closest to you that carries all the replacement parts for the system you choose. When you need a part for your system, you need it now. So make that your major consideration when buying a rack and pinion.

There are three different styles of steering racks – center pull, tie rod end, and threaded end.

The center pull style of rack is designed to place the tie rod loadings laterally into the rack rather than place an up and down loading on the rack and housing ends. This gives the unit a lot more strength and reduces wear. The one problem with this type of steering rack is that it lowers the unit about 3/4-inch. This means that the bottom of the rack housing has to be spaced upward this much to get it at the correct height for proper bump steer.

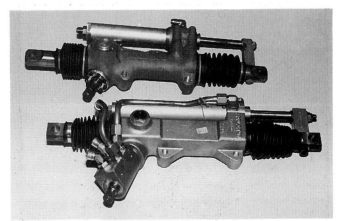

At top is a Woodward power steering rack and pinion. Bottom is an Appleton rack and pinion with power steering. Both of these units have a threaded end style of shaft.

The tie rod style of rack has a tapered hole in the end of the rack shaft to mount a standard tie rod end in it. The tie rod end is bolted in place on the top of the rack shaft ends, so there is no adjustment available there for bump steer. But the rack housing can be shimmed up and down to help with the adjustment.

The threaded end style of shaft attaches to a 5/8-inch spherical rod end instead of a tie rod. A 5/8-inch fine thread bolt captures the tie rod end and threads into the rack end. Spacers are used between the rod end ball and the rack end so that the ball doesn't bind during travel. The thickness of these spacers can be adjusted to help adjust bump steer. The rack can also be shimmed up and down to help in bump steer adjustment.

Rack and pinion units are available in a variety of ratios. They range from 16:1 to 8:1. Most paved track stock cars use a 12:1, 10:1, or 8:1 ratio. The 8:1 ratio is very quick and needs power steering for a driver to comfortably use it.

There are two different ways of referring to rack and pinion steering ratios. An 8:1 ratio is also referred to as 3-inch (or 3.4-inch in the Appleton line), the 10:1 as 2.88-inch (Woodward only), and 12:1 as 2.5-inch. The second number is the rack eye travel in inches per pinion turn. In other words, it is how far the rack travels with one turn of the pinion gear. One full pinion turn is also the same as one full steering wheel turn.

Because of the difference in various rack and pinion gears, and spindle steering arm lengths used, the ratios given here are only approximate. But you

(Left) The Appleton rack has the power steering servo attached to the rack itself which makes a compact installation. (Right) A Sweet power steering servo incorporated into the driveshaft.

can easily calculate the overall steering ratio for your car. With the front wheels setting on degreed wheel plates, turn the steering wheel one complete turn. Read how many degrees the right front wheel turned. Then divide 360° by this number. For example, if the right front turning angle was 30°, 360° divided by 30° equals 12, or a 12:1 overall steering ratio.

The standard steering ratio used for most short track cars is 12:1. It is the best compromise between speed, reaction and steering effort. The actual effort depends on tire width, car weight and scrub radius width. The wider the tire and the greater the scrub radius, and the heavier the car, the greater the steering effort. The slower the ratio, the less steering effort involved. A 16:1 ratio is slower than 12:1 and requires less driver effort than the 12:1. But a 16:1 ratio is too slow for the driver to keep up with the car on a short track.

The steering arm length makes a difference also. A 2.5-inch rack ratio is actually going to be a faster steering ratio on spindles with 5.5-inch long steering arms than on spindles with 6-inch long steering arms. All of these elements must be considered when choosing the rack and pinion ratio.

When using a 12:1 or faster ratio, power steering is a necessity. With 11-inch wide tires, a 2,600 to 3,000-pound car and 4° of positive caster, the steering effort would be too great for manual steering. Power steering is the only way to manage the steering effort involved.

Rack and Pinion Servicing

Periodically the rack and pinion unit should be disassembled, cleaned and greased. Be sure to check the pinion gear, rack, and side bushings for signs of wear. If the steering gets to the point where it feels sloppy to the driver, the preload should be tightened up. Refer to the instruction manual that came with your unit. Most rack and pinion units have a large adjusting screw that can be adjusted to tighten the preload or backlash. Be sure the backlash is not set too tight, or else the pinion gear will experience heavy wear.

If the rack and pinion unit is ever involved in a crash, it must be disassembled to check for internal damage. Check the rack to be sure it is straight. Check for chips in the rack. If there are any, it must be replaced. Also check for damage on the rack bushings.

Power Steering Pumps

A racing power steering pump requires a large pump capacity in both volume and pressure. Remember that a power steering pump and the servo have to be calibrated to work together.

For power steering fluid lines, use only steel braided Teflon core -6 (dash six) size lines for the pressure line to the servo. A high quality hose is required here because output pressures can exceed 2,000 PSI. Use a -10 (dash ten) line from the fluid reservoir to the pump.

Use only genuine GM or Sweet Manufacturing power steering fluid. Do not use automatic transmission fluid in your power steering pump. The power steering fluid has to be designed for extreme heat and have good anti-foaming qualities.

Excessive fluid temperature is the leading cause of power steering component failure. High engine RPM along with a large diameter drive pulley can cause a high fluid temperature. The steering system should have some type of provision for cooling the fluid. It is also advisable to use a filter in the power steering system. Fluid gets dirty and this can ruin seals and accelerate wear in the system. A power steering fluid cooler and filter are good options to consider. Mount an inline cooler between the pump and the servo. AFCO offers a combination finned fluid reservoir and filter.

Be sure a proper fluid level is always maintained, and change the fluid often. If the fluid turns dark or smells burnt, change it.

Installing Power Steering

Make sure all lines are flushed out thoroughly with clean power steering fluid before final installation. Debris in the lines can block the servo valve and damage the system.

With everything installed and the fluid filled, the system should be bled. Turn the steering wheel from lock to lock, and then hold it against the stop on one side to bleed. Always check the fluid level after bleeding.

Be sure to keep all power steering lines and the servo valve mounted outside of the driver's compartment. Because operating power steering fluid is extremely hot, this will keep some heat out of the driver's compartment, and most importantly, prevents burns to the driver should a line or seal fail.

The servo valve, which regulates the fluid delivery to the slave cylinder, is available in three different stiffnesses that regulate steering effort. Most brands are offered in light (easy to turn), medium (moderate turning effort), and heavy (harder to turn).

Designing And Building A Steering Shaft

The proper design of the steering shaft is critically important for the safety of the driver. The shaft has to be designed so that it and/or the steering wheel is not pushed back into the driver in case of a frontal impact. To protect against this, the shaft must incor-

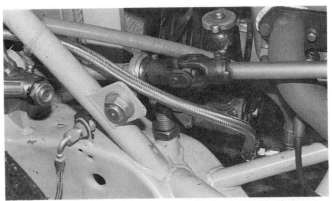

The first section of shaft is angled outward, then attached to another u-joint, which can be welded at each side, or have a splined connector to the second section of the shaft. Be sure that a spherical rod end bearing supports the steering shaft just forward of the u-joint connection.

porate a collapsible section, or should be built in multiple sections connected by universal joints that will fold up under impact. Or, a combination of these methods can be used.

Building A Multiple Section Shaft

The multiple section steering shaft is the most universally used, because in most race car applications there is not a straight shot from the steering wheel to the steering box. So, two sections of shaft are used, connected by universal joints, to bend around the interfering components.

When the angled sections of the steering shaft are laid out, they should be arranged so that when they are pushed from the front, the angled center section will push outward or, preferably, downward. This prevents the shaft from pushing backwards into the driver.

The steering shaft itself can be fabricated from .75-inch OD, .3125-inch ID DOM tubing. Mild steel tubing is lighter than solid stock.

A multi-section steering shaft is laid out as follows: a universal joint is attached to the rack and pinion unit. This u-joint is splined on one end to attach to the steering box, and has a smooth bore on the other end. The steering shaft tube is inserted into the smooth bore, and welded around the circumference of the u-joint/shaft intersection.

The first section of steering shaft is joined to the next by a smooth bore/smooth bore u-joint. The shafts to the u-joint are welded on both sides. There are now two u-joints in the system, which is the minimum.

The second section of steering shaft passes through the firewall and into the driver's compartment where a quick release hub is welded to the end of it. Where the shaft passes through the firewall, many racers use a self-aligning shaft bearing (available from AFCO and Speedway Motors). It gives extra support to the shaft and effectively covers the firewall hole required to install the steering system.

When a power steering system is used, generally a short section of shaft will go from the quick release hub to a u-joint on the other side of a firewall, then a short section goes from that u-joint into the power steering servo, then another short section of shaft from the servo to the second u-joint. This extra work can be eliminated if the servo is an integral part of the rack, such as the unit built by Appleton.

Be sure to use spherical rod end bearings as a support for the steering shaft. These support bearings should be placed just forward of each universal joint in the system. The bearings must be properly attached to the chassis structure so that the steering shaft cannot move the support. Keep the steering shaft support bearings mounted as far away from the headers as possible. High heat generated by the headers can damage the bearings and universal joints, causing them to bind or have a tight spot.

AFCO and Aurora offer a special spherical rod end bearing designed especially for use as a steering shaft support bearing. The bore of the bearing is 0.007-inch oversized (.757 ID) so the rod end slides right over 3/4-inch steering shaft material — no need to grind out a standard rod end bore.

After the steering system is laid out and the position of all components is determined, the universal joints can be welded to the steering shaft. The welding can be a very critical situation. Not only does the driver's safety and health depend on the quality and integrity of the weld, but welding heat can cause serious damage to the joint. Welding can distort the housing, or freeze up the pins on a needle bearing joint. To prevent this type of damage during the welding process, keep the universal joint center section cooled by wrapping a wet shop cloth around it. Once the weld is completed, let the u-joint and steering shaft cool in the ambient air. Do not use water to cool them. If you are not an experienced, competent welder, do not attempt to weld the steering shaft and u-joints yourself. Get the help and experience of someone who is.

TIG welding is preferable to MIG welding for attaching the u-joint to the steering shaft because you can get better penetration with better control of the heat. If you choose this method, be sure to have an expert in TIG welding do it.

Although universal joints are designed to operate at an angle, increasing the operating angle past 30 degrees significantly increases the stresses imposed on the joint. Do not do it. If your steering application requires operating a u-joint at close to 30 degrees, you should redesign the steering to employ another universal joint. It is always best — in terms of safety — to design a steering shaft system that employs at least two universal joints. When two universal joints are used in the steering layout, make sure the forks of each are installed in line with each other.

There are several different brands of universal joints to choose from. Choose one made by a company that manufactures universal joints specifically for racing and high performance use. These parts have been torque-tested to limits well beyond what a race car should endure and have survived.

Steering Wheels

Different steering wheel diameters create a different feel for the driver. The diameter choice strictly depends on what the driver feels most comfortable with. A larger diameter wheel requires more driver input for a given amount of front wheel directional change. A smaller diameter wheel is more precise – it requires a much smaller amount of driver input for the same amount of directional change. The steering wheel size can change the feel of a particular race car from that of a large diesel truck down to that of a precision sports car.

Manufacturers offer steel and aluminum steering wheels in 15-inch and 17-inch diameters with or without finger grips. Reb-Co offers a 16-inch diameter steering wheel.

The steering wheel dish depth can be used to help properly position the driver in the cockpit. Most steering wheels are offered in 1.5-inch or 3-inch dish. If you want to position the driver further back in the cockpit to improve rear weight distribution, using a 3-inch dish steering wheel can be advantageous.

Rear Suspension Systems & Driveline

On any type of track, the key to performance is getting good bite and acceleration from both rear tires. This involves an interaction of weight transfer, induced downforce loading of the tires, and the proper compliance at the tire contact patch. The relationship between the tire contact patch and the track surface can be a very fragile one when the tires are subjected to both lateral acceleration and forward thrust loadings. This relationship can easily be destroyed by a harsh torque reaction or a violent suspension movement. The spring rates and suspension attachment linkages must be carefully designed to cushion these harsh and violent reactions, and preserve the fragile contact patch/track relationship.

First we will discuss suspension theory that directly affects the rear suspension and the tuning of it. Then we will discuss all of the various rear suspension systems, their relative merits, how they work and how to tune them.

Pay attention to the rear suspension system. It determines how a car corners and how it is propelled forward. It is, in essence, the controlling element of the entire race car.

Rear Roll Center Location

The rear roll center location for a paved track stock car should fall between 8 and 12 inches above ground. The rear roll center location for a paved track car will depend on the type of rear suspension system used, the overall vehicle weight, how high the weight masses are placed in the chassis, the height of the front roll center, and the center of gravity height (CGH) of the car.

The vertical rear roll center is found on a car equipped with a Panhard bar at the distance from the ground to the center (halfway between the two mounting points) of the bar. On a car equipped with monoleaf or multileaf springs, the roll center is found at the point where the springs are anchored to the axle housing. In the case where a lowering block is used, the roll center is located half way between the top of the leaf spring and the bottom of the axle housing.

The rear roll center has its own "separate identity" from the front end. The front roll center must be established first because that one is the most critical, and the most difficult to change. Then work with the rear roll center. Make your best guess for rear roll center height, then do track testing to confirm your choice. Rear suspension systems that use a Panhard

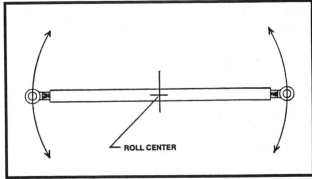

The Panhard bar length determines the lateral location of the rear roll center. The roll center is at the absolute center of the Panhard bar, regardless of whether that point falls on the vehicle centerline.

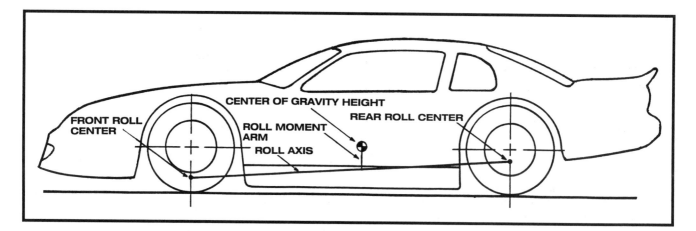

bar are more flexible in their use of the rear roll center as a tuning element. The attachment brackets can be designed so that one side of the bar, or both, can be easily raised or lowered to adjust the roll center height.

Rear Roll Center Lateral Location

The roll center of a suspension system is that point at which lateral forces can be applied without creating rolling of the sprung mass. It is, simply, the center of a rolling object. The center will remain still while the outsides will move in equal arcs. In coil-sprung stock cars that have a Panhard bar as a rear

lateral locating device, the length and mounting location of the Panhard bar determines the rear roll center. If, for example, the race car has a short Panhard bar which anchors to the chassis on the right and to the rear end housing at the center of the car, the roll center is not at the center of the vehicle but rather halfway between the attaching points of the Panhard bar. In this case, the rear roll center would actually be located about 12 to 15 inches to the right of the mechanical center line of the car. This would create a roll axis for the car which runs at an angle to the center line of the car (see drawing).

The rear roll center location with a solid rear axle depends not only on the geometry of its layout but

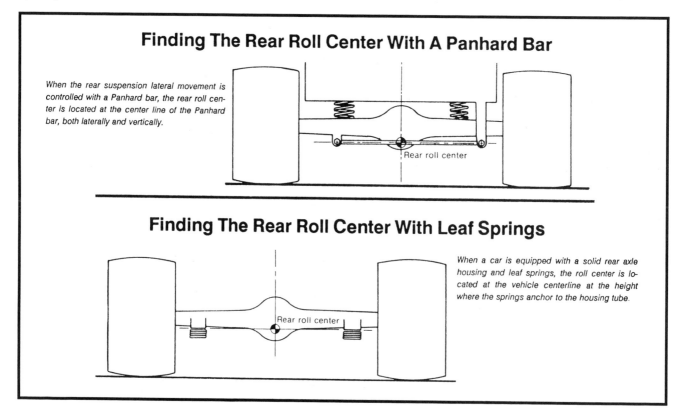

Finding The Rear Roll Center With A Panhard Bar

When the rear suspension lateral movement is controlled with a Panhard bar, the rear roll center is located at the center line of the Panhard bar, both laterally and vertically.

Rear roll center

Finding The Rear Roll Center With Leaf Springs

When a car is equipped with a solid rear axle housing and leaf springs, the roll center is located at the vehicle centerline at the height where the springs anchor to the housing tube.

Rear roll center

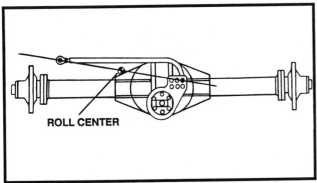

When a J-bar is used as the Panhard bar, the roll center is located at a point midway between the 2 mounting points.

also is influenced by relative spring rates. Having a stiffer spring on one side than the other will move the roll center toward the stiffer spring. How much does the unequal spring rate move the roll center? We calculated an example using a 60-inch rear track width, a 40-inch rear spring track (center of the left rear spring to center of the right rear spring), a 200 lb/in spring at the left rear and a 250 lb/inch spring at the right rear. The roll center was shifted to the right (toward the stiffer spring) 1.88 inches.

It is apparent, then, that if a short Panhard bar is used which offsets the roll center to the right, a stiffer left rear spring can be used to offset it back toward the center line of the car. Something on the order of at least a 50 lbs/inch stiffer spring at the left rear should be used with the short Panhard bar when it mounted on the right side.

What happens if the short length Panhard bar is attached to the left side of the frame instead of the right side? The roll center will be offset to the left of the vehicle center line, and it will now require a stiffer right rear spring than a left rear. If this were a high cross weight car, this would not be a desirable situation, because the higher left side weight would require a stiffer spring at the left corner. Having the Panhard bar attached to the chassis on the left side also means that the roll center rises as the body rolls. This loosens up the chassis more as the body rolls more.

Some Rules Of Thumb For Rear Roll Centers

1) A highly wedged car (high cross weight, in the area of 58 to 60 percent) along with a short Panhard bar mounted on the right will keep the rear of the car

very tight. The left rear spring should be stiffened to compensate for this.

2) Cars with less cross weight, in the range of 52 to 55 percent, perform better with the rear roll center in the center of the car or offset toward the left rear tire.

3) If you use stiff spring rates, you can run lower roll centers. Or, turning that around, if you have lower roll centers, you are going to need stiffer spring rates to control the body roll of the car.

4) Rear roll center location should be somewhere between the bottom of the ring gear (assuming the use of a Ford 9-inch rear end) and just below the axle housing tube. So the ballpark figure is somewhere between 10 and 11.5 inches. Be aware that tire height is going to influence the roll center height. If you bolt on a set of tires that have a lower section height, you will lower the roll center. A taller tire will raise the roll center on the same chassis. It is best that you have a Panhard bar (or other attaching linkage) that is adjustable to handle this.

5) Rear tire width has an influence on rear roll center height. When you have a small rear tire, the rear roll center should be lowered to get more bite. When you have large width rear tires, you want to get the rear roll center up higher, or else the car won't turn at the apex. With a low rear roll center and wide tires, you will have too much rear traction, and thus a chassis push.

The Roll Axis

The roll axis is an imaginary line connecting the front roll center with the rear roll center and is theoretically the axis about which the body rolls during cornering. The ideal chassis design would have the roll axis on the mechanical centerline of the car. This would make the car roll equally left to right about one axis line. To increase or decrease bite, then, you could play with left and right spring rates. There would be no need to add or subtract spring rate at one corner or the other to support extra weight leveraged from a displaced roll center.

Mass Axis Vs. Roll Axis

The mass axis can be pictured as a line running from front to rear, as seen from above, connecting the center of the major concentrations of weight in the vehicle. It would be like slicing the car into

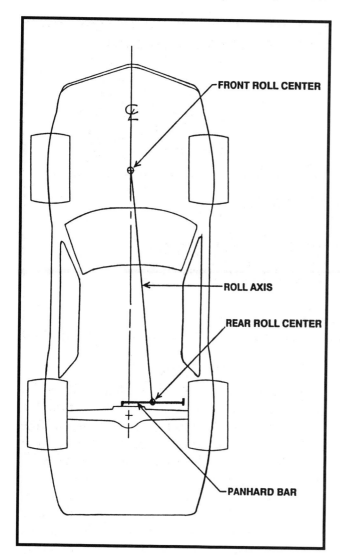

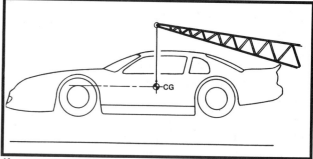

If you cut a hole in the roof of a car and positioned a crane so it could hook the center of gravity, it would suspend the car perfectly balanced.

and lower weight masses of the car. We concern ourselves with the center of gravity height because it is that point through which the centrifugal force acts during cornering. The higher the CGH is above the roll axis, the more weight transfer from inside to outside during cornering.

For a car running on a paved track, the CGH should be as low as absolutely possible to minimize the inside to outside weight transfer. This helps keep more bite at the left rear tire. For paved track stock cars, the CGH usually ranges between 15.25 and 17.5 inches, depending on engine height and ballast placement.

Determining the precise CGH of your car can be very difficult if you do not have access to wheel scales. If you want to closely approximate the CGH, use the following guidelines: for most cars, the measurement from the centerline of the camshaft to the ground is very close to the CGH figure. For very highly modified chassis built with the ultimate placement of weight in mind, and with a large amount of ballast placed very low in the chassis, this approximated CGH can be lowered by 2 to 2.5 inches.

Front Vs Rear Roll Center Height Relationship

The front roll center should always be lower than the rear roll center. If the front roll center is too much lower, the car will roll too much too quickly in relation to the rear at turn entry (most probably creating a push, depending on other factors). On the other hand, if the front roll center is right and the rear roll center is too high, the car is going to tend toward oversteer upon turn entry (with all other factors being equal). Wheelbase length is also a design consideration in roll center height relationship. The shorter the

several sections, like a slice of bread, and locating the left to right center of weight in each slice, then connecting all these centers with a line. This axis will not be a straight line, but can be projected into one to give an indication of its relationship to the mechanical car center line and roll axis.

Every car has its own identity and you have to adapt roll centers, mass placement and spring rates to what the car likes. Only testing will tell you for sure. The ideal chassis design would put the roll axis and the mass axis very close to each other.

Center Of Gravity Height

The center of gravity of a car is that imaginary point which is the absolute center of all weight in the car — vertical, front to rear, left to right.

The center of gravity height (CGH) is the balance point in the chassis which evenly splits the upper wheelbase, the more critical the front/rear roll center relationship is.

There is no set mathematical relationship between the front and rear roll center heights. The only way to design any race car is to use sensible ballpark guidelines for the front roll center as outlined before. The rear roll center should also be built within ballpark guidelines (between 10 and 12 inches). Then only actual testing will show you the ultimate placement for the rear roll center. The rear linkage which sets the roll center height should be moved in 1/2-inch increments, with a test session for each placement. As track testing gives you feedback about how different roll centers affect your car's handling, make notes about this for future fine tuning help. (In the Chassis Adjustment chapter, there is a discussion of using the rear roll center as a fine-tuning device.)

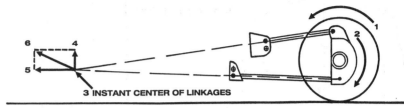

FORCES AT WORK DURING ACCELERATION

3 INSTANT CENTER OF LINKAGES

1. TIRE TRAVEL – COUNTERCLOCKWISE
2. REAR END HOUSING CLOCKWISE FORCE REACTION TO APPLIED TORQUE – CALLED WRAP-UP
3. THE INSTANT CENTER OF THE REAR END LINKAGES APPLIES THE FORCES TO THE CHASSIS – LIFTING AND PUSHING. TIRE THRUST IS PUSHING THE LINKAGE FORWARD WHILE TORQUE REACTION OF THE REAR END HOUSING IS APPLYING LIFTING FORCES AT THE INSTANT CENTER AS A RESULT OF HOUSING ROTATION.
4. LIFTING
5. FORWARD THRUST
6. RESULTANT APPLICATION OF FORCES

Front Vs. Rear Track Width

For a short track late model sportsman (both dirt and paved track applications), a slightly narrower rear track width creates a more stable handling car. The narrower rear track makes the car tighter both on turn entry and acceleration at turn exit. Typically, the front track width is set at the maximum allowed under the rules, and the rear track width is set 2 to 2.5 inches narrower than the front.

An independent front suspension also is improved by having a wider track because it provides a wider support base for the car.

When running on a track that is real fast and fairly highly banked where you keep up the momentum through the turns, you want to have a rear track width equal to the front. This is so the car doesn't bind up and scrub off speed. The easiest way to handle this is to change the right rear wheel to a different offset which moves the right rear out in line with the right front.

Anti-Squat

Anti-squat increases rear tire loading under acceleration. Squat is the weight transfer force from the front to the rear under acceleration. Anti-squat is the reaction of the rear suspension forces acting against the chassis. The result of anti-squat loading is increased tire traction.

Anti-squat is a very important element in a chassis. It helps to firmly plant the rear tires onto the track surface. As the vehicle accelerates, weight is pitched rearward in reaction to the vehicle's forward movement. This rearward pitch of vehicle weight can be reacted in two ways at the rear suspension. The first case is with no anti-squat. Here the weight is utilized simply by the springs using up compression travel. No additional weight is transferred downward onto the tires. The second case is a car with anti-squat. Anti-squat utilizes the linkage of the rear suspension to brace the rear end housing against the body moving downward on it. The linkage acts as a lever to apply more force to the rear end housing as the body squats. When rear tire loading is increased by the use of anti-squat, more traction is available at the rear tires. This means that the driver can use more throttle quicker at corner exit.

Anti-squat is expressed as a percentage. Fifty percent anti-squat means that 50 percent of the rearward pitching travel is reacted by the mechanical leverage of the rear suspension arms. One hundred percent anti-squat means that 100 percent of the

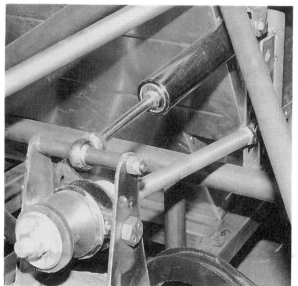

To prevent wheel hop under braking, an axle damper shock mounted above the rear end center section is used, angled uphill 5 to 7 degrees.

rearward pitching moment is reacted by the mechanical leverage. With 100 percent anti-squat, the rear of the vehicle will not squat at all. It is possible to obtain more than 100 percent anti-squat. In this case, the rear suspension linkage begins to multiply the mechanical leverage, and the body actually is lifted upward during forward acceleration.

Leaf spring suspensions offer far more than 100 percent anti-squat. Their anti-squat leverage is not delivered all at one time under hard acceleration, and this deformation of the springs cushions some of the harsh reactions. This is an ideal situation.

There is one drawback associated with the use of high amounts of anti-squat. That is rear wheel hop under heavy braking. The mechanical leverage that creates anti-squat under acceleration will lift the chassis away from the rear axle under deceleration. This causes the unloaded rear axle to chatter and hop. To prevent this from happening, keep the anti-squat under 50 percent and use an axle damper shock mounted above the rear end center section, angled uphill at 5 to 7 degrees. This controls the braking forces and prevents rear wheel hop by keeping a load on the rear end housing.

Axle hop can also be controlled by mounting the rear axle brake calipers to brake floaters. Brake floaters remove the braking torque forces from the rear suspension components and direct them into the chassis by use of a separate set of linkages. See more details on brake floaters in the Brakes chapter.

Rear Roll Steer

Roll steer is a change of steering angle in the rear end due to chassis roll. Roll steer is present if the rear axle turns (relative to the centerline of the chassis) during chassis roll. Another way to picture it is that the wheelbase grows slightly longer on one side of the car, and slightly shorter on the other side of the car. It doesn't take a very big steering angle of the axle to make a significant change in handling.

On a solid axle (either front or rear), roll steer is caused by geometric differences in the arcs of each wheel (on the same axle) during body roll due to linkage arrangements. One wheel is moved backward or forward relative to the other during body roll. This can happen because one side has a different length longitudinal control arm than the other side. Or, if the left and right side control arms are symmetrical (which is most often the case), it can happen because the arcs of the two arms are different during bump travel compared to rebound travel. Remember that during body roll, one wheel (the outside one) is in bump while the other is in rebound.

Rear roll steer can occur as either roll oversteer or roll understeer. If the outside wheelbase grows longer than the inside during body roll, this angling of the rear axle creates roll oversteer, because the axle is steering the rear of the car toward the outside of the turn. If the inside wheelbase of the car grows longer relative to the outside wheelbase during body roll, roll understeer is present because the rear axle of the car is steering the rear end to the inside of the track.

The existence of – and the degree of – roll steer is determined by the longitudinal (front to rear) rear suspension control arms, with a coil spring suspension. Their lengths, mounting points and their angle relative to each other and the ground plane are all factors.

With leaf spring rear suspension, the mounting angles of the springs determine the amount of or degree of roll steer. If the springs are not mounted parallel to each other in the longitudinal plane, they will scribe different arc movements at the housing centerline, and will create roll steer. This is why weight jacking devices or multiple mounting holes are not recommended for mounting leaf springs. The proper way to mount a leaf spring is with a slight downward angle, viewing the spring in a side view plane, with the front eye of the spring slightly lower

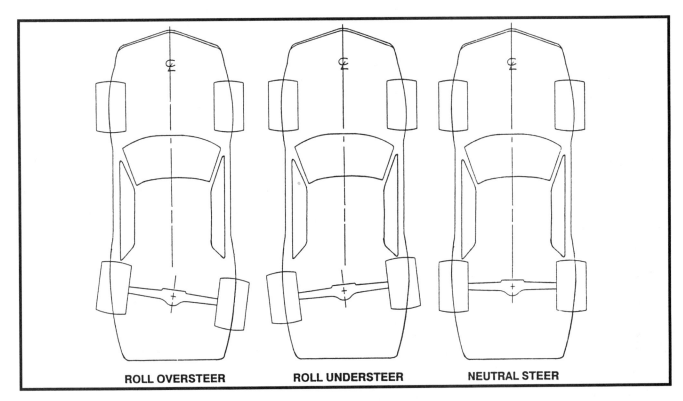

ROLL OVERSTEER   **ROLL UNDERSTEER**   **NEUTRAL STEER**

than the rear spring eye. The downward angle should be about 5 degrees. For most standard design leaf springs, this keeps the front section of the spring (which functions more like a control arm than a spring) parallel from side to side during normal chassis movement and body roll.

Using Roll Steer

First of all, measure your car through increments of bump and rebound travel on the inside and outside wheelbases, and determine the existence of roll steer in your car.

Roll understeer should be avoided. It is difficult to balance the chassis with it present. The only way to balance the chassis (without fixing the actual cause of the problem with the suspension linkage) is to increase rear oversteer by adding too much rear tire stagger. But, this is introducing a new problem into your chassis to cover up an existing one.

A small degree of roll oversteer can be helpful, especially with a car which has a solidly locked rear end. It works in a positive way the same as a small amount of tire stagger does. The rate of roll oversteer will be proportional to the rate of body roll. If a driver

uses the diamond pattern of driving, there will be only a small amount of body roll during initial turn entry and straight line braking. A higher cross weight (more understeer) helps keep the car stable here. Then when he makes his turn to the left at the apex, a sudden amount of body roll is induced, and with it comes a sudden amount of roll oversteer, which helps him cleanly make the turn. Then its a straight acceleration off the turn, with body roll rapidly decreasing.

Spring Loaded Radius Rod

In cases where a driver relies on skewing the rear end housing to create roll understeer, an alternate choice might be to use a spring-loaded radius rod on the right side of the axle housing (when a 3-link suspension is used). This compressible radius rod skews the rear end housing only in response to acceleration.

When a spring-loaded radius rod is used at the right rear, axle thrust under acceleration pushes forward against the link. The spring compresses and thus the right side wheelbase is shortened, creating understeer. This is used on a car that suffers from power oversteer. Once the heavy acceleration forces level off, the spring rate returns the link to its original

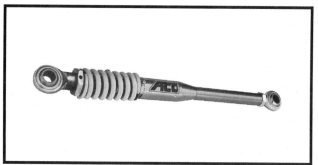

A compressible spring-loaded radius rod used at the right rear creates roll understeer under acceleration.

length and it operates normally just like a solid radius rod.

The amount of spring rod compression depends on the rate of the spring used. A stiffer spring resists more thrust and thus compresses less.

The amount of spring rod compression can also be controlled by preloading the spring. Preloading stiffens the coil spring's rate. If 1/2-inch of preload is added to the spring, a much higher amount of thrust force must be overcome before the rod's spring compresses. For example, if a 1000 lbs./inch spring is preloaded 1/4-inch, a force of 250 pounds is fed into the linkage. This means a thrust force of more than 250 pounds must be encountered before the spring rod will begin compressing. The preload value can be adjusted to tune the suspension link to the amount of spring compression desired to overcome power oversteer. Spring rates can also be stiffened or softened to adjust the amount of axle skew.

Spring rod travel should using not exceed 1/4-inch. If it has to in order to balance full throttle oversteer, then the chassis spring rates must be changed to balance the chassis.

Rear Suspension Attaching Linkages

There are four different movements that the rear end can experience: side-to-side, fore and aft (front to rear), pitch about its axis (rolling forward or rearward), and rotation about a vertical axis (a vertical axis is one up and down through the center section, and the rotation is one which is seen from looking down from above).

Because there are four different directional movements that the rear axle can experience, the maximum number of linkages which can be attached to

it is four. If any more linkages than that are used, the axle is subject to mechanical bind as it goes through its movements.

The rear axle restraints for a coil-sprung car include two trailing links (which control fore and aft movement and rotation about a vertical axis), an upper trailing link (which controls pitch about the axle's axis), and a Panhard bar (which controls lateral or side-to-side movement). This is a description of the basic three-link rear suspension system.

3-Link Rear Suspension System

The three-point rear suspension linkage has the most wide-spread usage for all stock cars employing coil or coil-over springs at the rear on paved tracks. The traditional three-point linkage includes two lower links attached one at each outer end of the rear end housing from a bracket on the housing straight forward to a bracket on the chassis. The third link is attached at the center on top of the rear end housing, running forward to a bracket on the chassis. The forward mounting height of this upper link can be adjusted for the amount of anti-squat in the rear suspension. The three-point rear linkage is very easy to work with and is highly adjustable so that it can be adapted to almost any type of racing condition.

The bottom links should be 20 to 24 inches in length running parallel to the ground. If your chassis can accommodate a 24-inch lower link, by all means use that length. Longer is better because it produces less angular change on the rear end housing as one side moves in bump and the other moves in rebound. It is most practical to keep the lower trailing arm lengths the same length on both sides. It simplifies things a great deal when you can carry one spare link and it will fit the left or the right side.

The upper link should be 18 to 20 inches in length mounted downhill (front end of it lower) at about a 5 to 7-degree angle. The length of the upper third link determines the maximum downhill angle it can operate at. A longer link allows slightly more downhill angle to be used. This mounting angle can be moved upward or downward 3 degrees in either direction to help serve as a fine-tuning device to add or subtract traction under acceleration. More on that later.

You should avoid any lateral angles with the lower links (pointing the links in or out as they run forward). These lower links are what propel the mass of the

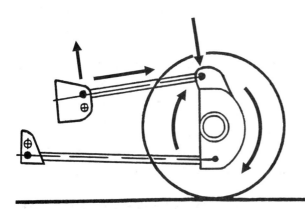

Reaction force from rear end thrust lifts up on the forward pivot point of the third link. The resulting reaction places a downward loading on the rear tires. The greater the downhill angle, the more lifting applied at the forward point of the upper link, and thus greater tire loading.

The mounting angles for the third (upper) link and axle damper shock are important for resisting forces in the correct direction. The upward angle of the axle damper shock helps tighten up the car under braking. The downward mounting angle of the third link provides for more anti-squat under acceleration.

car forward from the rear tires. They push the race car around the track. If the links run at various angles, it dissipates the energy being used to push the car forward. The lower links should push the car forward without binds. The upper link, as well, should always be mounted straight and parallel to the vehicle centerline. This prevents any binding of the linkage under acceleration and deceleration.

The geometry – or angles – of a three-point link system, projected forward to a common point, form an instant center, and thereby behaves like a torque arm. Because of this, be very careful in the layout of the three-point rear linkages so that this instant center doesn't have a lot of movement during rear suspension travel. Having very steep angles of the upper and lower arms will create a short instant center, causing it to move around. If the suspension travel causes instant center movement, the rear wheels and suspension arms will be changing angles quite rapidly, making for a very unstable feeling for the driver.

The angle of the upper link in a three-point system has a very important influence on the performance of the chassis under braking and acceleration. If the upper link is mounted at a downhill angle toward the front, this will promote more anti-squat under acceleration. This means the rearward pitching moment of the car under acceleration is reacted by the mechanical leverage of the suspension arms and thus causes additional downforce to be placed on the rear tires. However, the big drawback of this is the opposite and equal reaction under braking. A

large amount of anti-squat (downhill angle of the upper link) will cause rear end lightness and rear wheel hop when the loads are transferred the opposite way under braking. The rear end will get very light, causing a lack of directional control and wheel hop under braking. The more downhill angle of the upper link, the more severe this problem will be.

An uphill angle of the upper link will promote good firm traction of the rear wheels under braking. But it has a big drawback under acceleration — it promotes pro-squat, which lifts the rear wheels up under acceleration and diminishes traction.

The answer is a compromise. The upper link should angled downhill at a 5 to 7-degree angle. This promotes good traction without the associated problems of rear end lightness and wheel hop. Additionally, an axle damper shock –mounted at an uphill angle – should be used in conjunction with the upper link.

The mounting angles for the third (upper) link and axle damper shock are important for resisting forces in the correct direction during acceleration and deceleration. The upward angle of the axle damper shock helps tighten up the car under braking during corner entry. The downward mounting angle of the third link provides for more anti-squat under acceleration.

Braking forces are controlled by the axle damper shock absorber which is mounted adjacent to or above the third link and is angled upward at a 5-degree angle. This type of shock absorber has 90 percent of its damping force in compression, and only 10 percent in rebound. A typical axle damper shock has 600 pounds of damping force in compression and only 65 pounds of force in rebound. So, the axle damper offers almost all of its resistance during braking, essentially acting like a cushioned solid link. But under acceleration, it offers almost no resistance and allows the solid third link mounted at a downhill angle to operate properly to provide anti-squat.

The upper third link should be mounted to the chassis at the center of the weight mass of the car. The center of the weight mass is found by multiplying the rear track width by the car's left side weight percentage. If a car has a 60-inch rear track width, and a 58 percent left weight percentage, the calculation is: 60 x .58 = 34.8. So, the center of the weight mass at the rear is located 34.8 inches to the left of the center of the right rear tire.

If the link is not mounted at the center of the weight mass of the car, under acceleration the loading of the rear tires will not be equal. It will add more force to the side of the car that the link is closest to. For example, if the link is mounted 10 inches to the right of the weight center, under acceleration it would add more loading to the right rear tire and unload the left rear tire by an equal amount.

The third link and the axle damper can be mounted side by side, or with the axle damper above the third link. The third link should be mounted 12 inches above the center of the housing axle tube.

The rear mount of each lower trailing link should be located within 8 to 12 inches of the brake rotor rear face. If the mounts are too far inboard to the center of the car, the leverage forces are going to be real hard on the rod ends' life.

Be sure to check the trailing arm rod end bearings for wear or freeze-up. These problems can create difficult-to-diagnose handling problems. Check rod end freedom weekly and keep them cleaned and lubricated.

Some Guidelines For Third Link Angle Setting

1) The initial angle for the third (upper) link should be 7 degrees downhill toward the front mount. No more than a 10-degree downhill angle should ever be required with an 18 to 20-inch long upper link.

2) Adding more downhill angle will tighten up the chassis more under acceleration off the corners by adding more tire loading. But that will also cause more rear end looseness and wheel hop under braking.

3) To control rear end looseness and wheel hop under braking, use an axle damper shock mounted at 5 degrees uphill. If rear end looseness persists at corner entry, use more axle damper uphill angle, adding no more than 2 degrees each time. 8 to 9 degrees should be the maximum uphill angle used.

4) If the rear end is too tight at corner entry, decrease the uphill angle of the axle damper shock. The adjustment range of the axle damper is 0 to 7 degrees uphill. More angle tightens the chassis at corner entry; less angle loosens the chassis at corner entry. Don't ever mount the axle damper with a downhill angle.

Spring Loaded Torque Link

The spring loaded torque link is an upper link mounted above the center of the rear end housing. It is a replacement for a steel tubular upper third link. The spring loaded torque link is installed in conjunction with an axle damper shock absorber, as described above, to control brake reactions.

The spring loaded torque link is a rod connected to a spring contained in a housing. The spring is compressed under acceleration torque to dampen the torque reaction at the rear tire contact patches. This provides a smoother application of power at the rear tires at corner exit and enhances tire life.

A torque absorbing link in the upper third link position can make a big difference in handling, especially when a "track tire" is required. With a hard tire compound, the torque reaction will break the rear tires loose much quicker and easier than with tires that are softer and more forgiving. The torque absorbing upper link can help cushion the torque reaction at the rear tire contact patches.

The torque link should be between 18 and 22 inches long. A longer length helps the suspension to

The spring loaded torque link compresses under acceleration torque to dampen the torque reaction at the rear tire contact patches.

work better. A shorter link reacts too quickly. The longer link creates a smoother arc as it travels while reacting against chassis squat. A smoother transition doesn't shock the rear tires as much during sudden application of power.

If a torque link is substituted for an existing solid third link, the forward mounting point is already fixed on the chassis, so usually a torque link the same length as the third link would be used. Most chassis builders use an 18 to 20-inch third link. This length is within the range of acceptability, but a longer torque link can be used if a rearward offset housing bracket is used for the torque link. This can add up to 3 inches of length.

The paved track torque link should be of a "single pull" design. "Single pull" means that it has a single spring inside the unit. This limits spring travel and yet provides cushioning at the rear tire contact patches. For paved racks, use the single pull model which is equipped with a rubber bushing end for better control under braking. The rubber bushing keeps the rear end of the car tighter as opposed to a unit which uses a brake reaction spring.

The torque link uses spring rates of 1,600, 2,000 or 2,500 lbs. per inch, depending upon the type of engine used in the race car, and the type of tires used. A car with a 9 to 1 compression engine, or restricted carburetion, or a car which uses harder spec tires or narrow tires would require a 1,600 lbs. per inch spring in the torque link. If the car uses wider or stickier tires, or an unrestricted engine, a 2,000 lbs. per inch spring would be used. Only if the car had

tremendous power and super sticky tires would a 2,500 lbs. per inch spring be required.

The proper spring rate is determined by the amount of spring travel in the link under acceleration. A 1/2-inch travel of the link under acceleration is considered ideal. The spring should be preloaded 1/2-inch. If less than 1/2-inch of travel is experienced, a softer spring rate should be used, or less spring preload should be used. If 3/4 to 1-inch of travel is experienced, a stiffer spring rate or more spring preload should be used. In both cases, adjust the spring preload first before changing spring rates.

The torque link should be mounted to the chassis at the center of the weight mass of the car, just as described previously for the third link connection.

Mount the torque link 12 to 15 inches above the center of the housing axle tube, and 1 to 3 inches behind the vertical center of the axle tube. The torque link and the axle damper shock can be mounted side by side or with the axle damper above the torque link.

The torque link is used as a tuning device by changing its mounting angle. The starting angle of the torque link is 6 degrees downhill toward the front mounting point. This angle can be varied by up to 3 degrees up or down for fine tuning. Adding more downhill angle adds more anti-squat, which tightens up the chassis under acceleration. Using less downhill angle loosens the chassis under acceleration at turn exit. More downhill angle loosens up the car at turn entry.

Be sure to use an axle damper shock in conjunction with the torque link to combat rear end looseness under braking. The axle damper should be mounted at 5 degrees uphill to start. If the rear end feels loose at turn entry or the car experiences wheel hop under braking, more uphill angle can be added. 8 degrees should be the maximum that a chassis should require. The chassis handling reactions should be sensitive to 1 or 2 degrees of angle change with both the torque link and the axle damper.

When fabricating a front mounting bracket for the torque link for multiple angle positions, make sure that the holes are drilled in a radius from the center of the housing mount of the link. This makes every mounting hole an equal distance from the back mounting point of the link.

When using a spring loaded torque link, the static rear end pinion angle should be 5 to 6 degrees

The rubber bushing torque link is typically used on IMCA modifieds and similar types of cars to cushion the torque reaction at the rear tire contact patches.

downhill. This is because the torque link will allows the pinion to move up 3 degrees under acceleration due to link compression. This setting prevents the pinion angle from going to 0 degrees or over center.

Rubber Bushing Torque Link

With IMCA modified and similar classes, a torque absorbing upper link is required to help cushion the torque reaction at the rear tire contact patches.With a hard tire compound, the torque reaction will break the rear tires loose much quicker and easier than with tires that are more forgiving.The rubber bushing torque link absorbs the energy shocks the suspension transfers to the rear tires under acceleration, which is very important with the narrow, hard spec tires used.

The rubber bushings on the torque link are available in three compression ratings – standard, medium and soft. The rubber bushings used for a paved track application have to be all standard hardness bushings. The medium and soft bushings are meant only for dirt track application and would create too much link travel on paved tracks.

Adding preload to the rubber bushings diminishes link travel. If 1-inch of preload is added to a link with standard hardness bushings, 400 pounds of compression force is added to the flexible portion of the link. Then movement in the chassis would have to exert greater than 400 pounds of force on the upper link before it started to compress. The chassis forces have to overcome the preload adjustment before any travel in the link occurs.

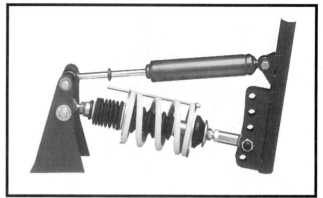

The AFCO IMCA-type torque link is especially designed for use with IMCA modified tires and engines. This link uses an external 5-inch OD spring.

The rate of the flexible portion of the link is varied by changing the preload on the rubber bushings. The ideal travel length is 1.5 inches. To achieve less travel with the link, increase the preload. To get more travel, decrease the preload.

The rubber bushing torque link has to be mounted laterally at the center of the weight mass of the car, just as described previously for the third link connection. Doing this allows both rear tires to be loaded equally under acceleration as the torque rod reacts against the rearward weight transfer in the chassis.

The rubber bushing torque link should be mounted 15 inches above the centerline of the rear end housing tubes using the standard 20-inch long link. It should also be mounted 3 inches behind the vertical centerline of the axle housing. The rearward offset makes room for a 20-inch long bar, and it is also less reactive on the rear end housing when weight is lifted off of the link at corner entry. The initial downhill angle of the link should be set at 10 degrees.

A more efficient type of cushioned upper link for IMCA modified and similar classes is the AFCO IMCA-type torque link.

AFCO IMCA-type Torque Link

The AFCO IMCA-type torque link is similar to a regular spring loaded torque link except that it is especially designed for use with IMCA modified tires and engines. This link uses an external 5-inch OD spring. A full range of spring rates are available, but the 900#/" spring (the stiffest available) should be used for paved track applications.

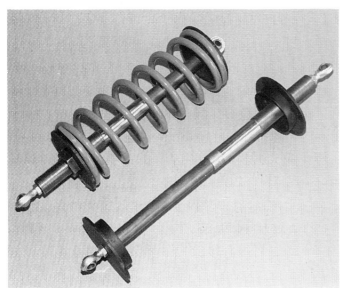

The coil spring slider is the most efficient way to mount coil springs in the rear suspension when coil-overs cannot be used. The top retainer of the slider unit rides on a threaded shaft and can be turned up or down to jack weight.

The mounting guidelines, desired travel and adjustment of this torque link are exactly the same as the spring loaded torque link.

Coil Spring Sliders

Coil spring sliders (also called coil-over eliminators) are used in IMCA modified and similar classes to mount 5-inch diameter springs to the rear suspension. These units are simple to use, are lightweight, attach easily to simple brackets on the chassis, and allow the springs to be mounted very close to the wheels. Coil spring sliders provide a weight jacking adjustment just like a coil-over unit. The top spring retainer of the slider unit rides on a threaded shaft and can be turned up or down to jack weight or set ride height.

If you use a coil spring slider, you must be aware that they are very critical on proper maintenance. These units use some close tolerance parts. Therefore they need to be disassembled often (every 3 to 5 race nights), inspected, and re-lubricated. Inspect for binds, galling of the shaft or a bent shaft. When reassembling, use anti-seize on the piston and light oil on the rod and bore. Dust and dirt can cause severe wear and/or binding problems on the slider, so be sure to check carefully and often.

Be sure that the slider unit you choose has some type of internal stop or retainer which prevents the

The Panhard bar height (at its center) determines the rear roll center height. The rear roll center height determines how much body roll a car experiences for a given amount of lateral acceleration.

spring from being dislodged when weight is taken off of the spring.

When mounting a spring with a coil spring slider, it is extremely important to make sure that the coil slider and shock absorber are mounted parallel to each other in all aspects. Both the coil slider and the shock are operating on shafts. so they have to operate in parallel. If they don't, it will bind up the suspension action.

The coil spring and slider unit should be mounted behind the rear axle housing, and the shock absorber in front of it, on both sides. This increases the spring base of the chassis. However, this arrangement can be changed to fine tune the handling reactions of the chassis. If the left rear spring and slider is mounted in front of the housing and the shock behind it, this will tighten up the chassis under acceleration, but loosen it up at corner entry. If the right rear spring and slider is mounted in front of the axle housing and the shock behind it, the chassis will be looser under acceleration at corner exit, but tighter at corner entry.

The Panhard Bar

The Panhard bar (or track bar as it is sometimes called) is a lateral locating linkage which connects the rear end housing at one side of the car to the chassis on the opposite side of the car. It is simply a tube with spherical rod end bearings at each end which braces one side of the car against the other to

This type of a bracket is one way of overcoming the clearance problem presented by a quick change when a full length Panhard bar is used.

prevent side shift of the chassis over the axle housing.

The Panhard bar height (at its center) determines the rear roll center height. The roll center lies at the center of the bar, no matter how long the bar is or where it is attached to the axle.

The rear roll center height determines how much body roll a car experiences for a given amount of lateral acceleration. A higher roll center decreases chassis roll because the lever arm distance between the roll center and the center of gravity height is shorter. A lower rear roll center generates more chassis roll because the lever arm distance is increased.

The lower rear roll center (and thus larger amount of chassis roll) will add more loading to the right rear tire during cornering. The weight will be loaded in a more vertical direction on the outside tire with a lower roll center. This creates more bite on the right rear tire, and tightens up the chassis.

Raising the rear roll center reduces body roll. That also makes the transferred weight load in a more angular manner at the right rear tire contact patch. This makes the right rear tire tend to slide laterally across the track surface. So, raising the Panhard bar – and thus the rear roll center – loosens up the chassis.

The full length Panhard bar is the simplest type of lateral control linkage. The full length bar will locate the rear roll center at the car's centerline, which

helps to minimize bump steer during body roll. When a Ford 9-inch rear end center section is used, clearance problems for mounting the bar behind the housing are not critical, and a full length bar can be used.

When a quick change center section is used, it presents a multitude of clearance problems for mounting a Panhard bar. To solve this, a shorter J-bar is used which attaches to a pinion mount bracket in front of the housing, and mounts to the chassis on the right side.

When a J-bar is used, the bar is level when the rod ends on each side are level with each other. If the J-bar is mounted with the longer portion of the bar level with the ground, there is an angle introduced into the mounting because the outside rod end is higher than the other end.

When a short J-bar is used, bump steer is introduced into the rear end. This happens because the arc described by the short bar is sharper than the arc described by the rear end housing as the car rolls. The other drawback of the short Panhard bar occurs in one-wheel bump along a straightaway. As a bump is encountered and one wheel moves up and down, the difference in arcs between the Panhard bar and the axle causes a lateral displacement of the tires. This means the rear tires are pulled left and right across the track surface. This scuffing will cause excessive tire heat and tire wear.

Panhard Bar Angle

The shorter J-bar type of Panhard bar has more of an effect on chassis handling reactions than a full length bar. For this reason, fine tuning of angles and mounting locations makes the chassis more reactive to this. To minimize erratic effects on handling, the J-bar length should be at least 20 inches long.

The Panhard bar ties the chassis to the rear end housing and tires. During cornering, this arrangement transmits a lateral force from the chassis to the tires. If the Panhard bar is level, this force is transmitted totally sideways at the rear tire contact patches. If the Panhard bar is angled uphill toward the right rear, the force transmission creates an upward force on the rear tires, which decreases traction on the rear tires.

The angle of the Panhard bar as well as its height can be used to fine tune how weight is transferred onto the tires at the rear of the car. However, exceed-

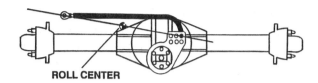

ROLL CENTER

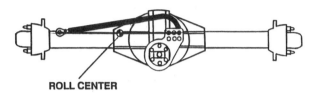

ROLL CENTER

If the Panhard bar is level, cornering force is transmitted totally sideways at the rear tire contact patches. If the Panhard bar is angled uphill, the force transmission creates an upward force on the rear tires.

ing reasonable limits of Panhard bar angle can create erratic changes in the rear tire loading. On paved track cars, a 1/2-inch change in height on one side of the Panhard bar can make a noticeable difference in handling characteristics. Start out with the Panhard bar mounted level at the desired roll center height. Then track testing will help determine the proper roll center height.

Panhard Bar Mounting Location

Forces that are transmitted through the J-bar type of Panhard bar are reacted on the axle housing at the point where the bar mounts. This affects how the forces split the loading between the left rear and right rear tires. If the lateral mounting location of the bar is at the center of the housing's track width, such as on a pinion mount bracket, each rear tire is loaded or unloaded equally. Thus, bar height and angle

An ideal way to mount a Panhard bar to the chassis is with a clevis-type clamp bracket.

changes will affect the left rear and right rear tires equally. The chassis mount side of the bar can be moved upward to loosen the chassis, or down to tighten up the chassis.

When the Panhard bar is not attached at the center of the rear track width, the loading caused by angle and height adjustments will not load the left rear and right rear tires evenly. If the axle housing mount is offset more toward the left rear tire, the left rear will be affected more by angle and height adjustments than the right rear will be. For this reason, it is always best to mount a J-bar at the center of the car's rear track width. The mounting position can be displaced an inch or two to either side of the true centerline and it will not create a noticeable change in handling.

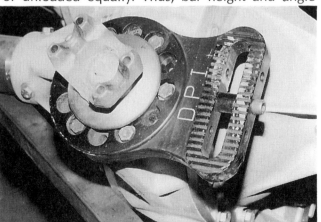

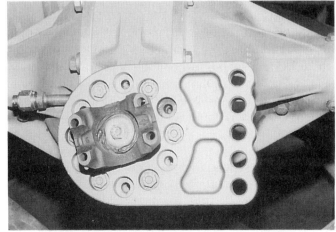

Two different styles of quick change Panhard bar mounting brackets. The one at the left has a serrated slide adjustment in 1/2-inch increments. The one at the right uses conventional threaded bolt holes.

An ideal way to mount the Panhard bar to the chassis is with a clevis-type clamp bracket. The bracket is a clamp-on mount which attaches to a round tubular upright chassis member. The end of the bracket is a clevis which serves as a straddle mount to capture the rod end bearing of the Panhard bar. There are two advantages to this type of mount. First, mounting to a steel round tube provides a lot more strength. Second, the clamp-on mounting bracket can easily be moved up or down to provide quick rear roll center height adjustments.

There are also two different types of quick change pinion mount brackets available. One uses conventional threaded bolt holes. The other attaches the Panhard bar to a serrated slide adjustment, which provides for height adjustments in smaller increments.

Left Vs. Right-Mounted Panhard Bar

When the outside mount of the Panhard bar is attached to the chassis on the right side, and the axle on the left, the rear roll center height lowers as the body rolls to the outside because the chassis mount lowers.

When the Panhard bar is mounted to the chassis on the left and the axle on the right, the rear roll center height rises as the body rolls to the outside because the chassis mount is raised.

The lowering roll center results in less lateral shear force at the right rear tire contact patch. This means more weight is planted on it. A rising roll center increases lateral shear force at the right rear tire contact patch, which decreases traction. A higher rear roll center loosens up the rear of a car, while a lower rear roll center tightens up the rear of a car.

There is something more important that happens when a short J-bar is mounted to the chassis on one side and an axle housing bracket. When the Panhard bar is mounted to the chassis on the right and the axle housing on the left, the rear roll center is displaced to the right of vehicle centerline (halfway between the two mounts of the Panhard bar). That means the entire roll axis is shifted to the right, and there is a shorter distance between the roll axis and the right rear tire than if a full length bar is used.

As the chassis rolls about the roll axis to the right, there is more weight transfer to the right rear tire with a J-bar because the roll center is moving downward a greater distance due to its offset from the centerline of the car. That loads the right rear tire and sticks the car tighter on corner entry.

Some Guidelines For Panhard Bar Adjustment

1) Lowering the Panhard bar equally (both ends the same) will tighten up the chassis during cornering.

2) Raising the Panhard bar equally (both ends the same) will loosen the chassis during cornering.

3) Increasing the angle of the bar (right side chassis mount moved higher) decreases traction to the rear tires, and loosens the chassis.

4) Decreasing the angle of the bar (right side chassis mount moved lower) increases traction to the rear tires, and tightens up the chassis.

5) If the axle housing mount of the J-bar is offset toward the right rear, this will result in more of the loading forces affecting the right rear tire, and less at the left rear. In this case, an increased loading on the right rear will tighten up the chassis at turn entry, but will tend to make the chassis loose at turn exit.

Clamp-On Mounting Brackets

Clamp-on brackets are available for the attachment of rear suspension components such as coil-overs, shocks, radius rods and other linkages onto 3-inch diameter axle housing tubes. Clamp-on brackets offer a big advantage because they are not welded to the rear end housing. This offers four major advantages.

First, it prevents rear end housing tube distortion caused by welding brackets. When brackets are welded to the tubes, warpage occurs. When a housing is fabricated by a professional builder, the housing is straightened after the brackets are welded on. When a racer welds on a housing in his shop, it is difficult to straighten the housing without the proper fixtures and skills. If a housing tube is improperly straightened, it could easily bend when it takes a hit.

The second advantage of clamp-on brackets occurs when the housing has been severely bent. If the mounting brackets are welded on, they most often get tossed out with the bent housing. Clamp-on brackets can be easily dismounted from the old housing and clamped onto the new one. And if bracket alignment or location becomes a problem

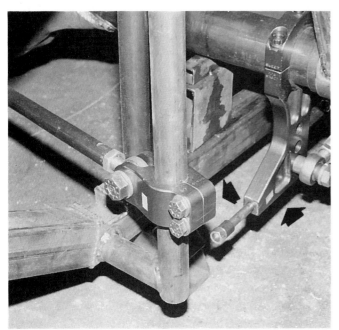

Clamp-on brackets are available for the attachment of coil-overs, shocks, radius rods and other linkages onto 3-inch diameter axle housing tubes.

when the new ones are installed, they can be un-bolted and relocated. This eliminates having to weld new brackets on the housing in a professional welder's jig.

Clamp brackets also allow a change in rear end tracking. The housing can be moved to the left or right simply by unclamping the brackets and shifting the housing. This allows a change in handling. When the housing is shifted left, it tightens up the chassis getting into a turn. If it is shifted to the right, it loosens up the car entering a turn.

Fourth, the clamp-on brackets can be used for rotation in adjusting pinion angle. The brackets can be loosened on the axle tubes, and the housing is then rotated to achieve the desired static pinion angle. With welded housing brackets, the brackets would have to be cut loose and rewelded.

When the clamp-on brackets are under heavy stress, they must receive additional help to keep them from rotating, especially if the housing brackets are resisting braking forces. There are two ways of preventing rotation. The first is to add a slight tack weld between the bracket body and the housing tube. This may be about 1/2-inch long. But the weld is only a superficial tack weld – not a deep penetrating weld that warps the housing.

The second way of positive attachment is to drill into the housing tube and use a set bolt inserted into

the indentation. This is probably the easiest way of preventing the brackets from spinning on the housing.

Other types of clamp-on brackets can be used to easily position adjustable linkages to the chassis. Chassis mounts for changing the height of adjustable linkages, such as trailing arm links and the Panhard bar, are most effectively mounted with clamp-on brackets. This allows the racer to make small adjustments instead of being forced to use brackets with holes spaced 1-inch on center. With clamp bracket mounting the range of adjustability is much finer and more precise.

Clamp-on brackets can also be used for mounting ballast. They give a racer the flexibility of mounting the ballast virtually anywhere on the chassis. Clamp-on brackets are available from AFCO to mount on both round and square tubes in a variety of sizes.

Coil-Over Mounting

Rear coil-overs should be mounted at an angle of 13 to 20 degrees inward to the chassis. This angle comes closest to pushing the shock shaft straight down, and upward again, during body roll.

Rear Shock Absorber Mounting Heights

How the chassis (or sprung mass) is connected to the suspension will make a big difference in how the car will react dynamically as weight is transferred side-to-side and front-to-rear. If the shock absorbers or coil-over units are mounted too high on the chassis, they will actually resist weight transfer.

The top of the coil-over or shock mount should be located as close to or below the CGH as possible. Ideally, it would be located below the CGH. But this would require shorter than 9-inch stroke shocks, and that is not an ideal situation for most cars.

To achieve a lower top locating point for the shock, the bottom locating point should be 7 inches below the rear axle centerline. That point is an ideal AND a maximum mounting point because placing the shock any lower many endanger the shock mount. This would occur if a car lost a rear tire and had to limp back to the pits on a rim. In most cases, the rim is located 7.5 inches below the centerline of the axle, so locating the shock mount at 7 inches below axle center leaves a .5-inch margin of safety.

Ideally, always try to mount the shocks or coil-overs 7 inches below the rear axle tube centerline.

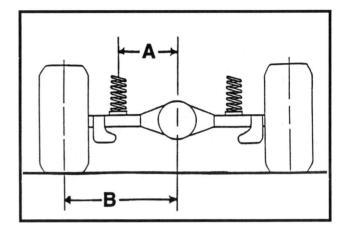

Locating the top mount of the coil-over or shock as low as possible enhances forward bite under acceleration. What happens when a car accelerates off a turn is that the G forces act against the CGH to transfer weight from front to rear. That weight transfer is resisted by the spring/shock upper mount. If the mount is 4 or 5 inches higher than the CGH, the transferred weight is going to just load the suspension links of the car instead of placing it on the springs and shocks where it can be used for forward traction.

However, if the top mount is located at or just slightly above the CGH, when the weight transfers under acceleration it can easily be transferred back and used for more forward traction.

The same concept is also valid for cars with conventional springs. If the shock absorbers are mounted too high in the chassis, they will resist the weight transfer onto the springs.

Ideally, always try to mount the shocks or coil-overs 7 inches below the rear axle tube center.

Rear Axle Suspension Motion Ratio

The motion ratio is the leverage created by a suspension linkage compressing a spring. The motion ratio multiplied by the spring rate – called the wheel rate – is the effective rate of the spring applied at the center of the adjacent wheel. The wheel rate is always lower than the spring rate because of the leverage effect of the linkage involved. The spring rate and wheel rate would only be the same if the spring was placed right at the center of the wheel.

The motion ratio for a straight axle suspension is found by dividing distance A by distance B (see accompanying drawing). A is the distance from the center of the spring mount to the centerline of the car, and B is the distance from the center of the tire to the centerline of the car. A can also be the center-to-center distance between the two suspension springs, and B can be the center-to-center distance between the two tires IF the chassis is not offset and the springs are attached to the rear axle in the same position on both sides of the car.

As an example, A = 22 inches and B = 32 inches:

$$MR = A/B$$
$$MR = 22/32$$
$$MR = .688$$

Wheel Rate

The wheel rate is the spring rate of the spring multiplied by the square of the motion ratio of the suspension attached to the spring. The formula is $WR = (MR)^2 \times SR$ where MR is the motion ratio and SR is the spring rate.

As an example, MR = .688 and SR = 275:

$$WR = (MR)^2 \times SR$$
$$WR = .688^2 \times 275$$
$$WR = 130$$

The Coil-Over Angle Factor

If the spring attached to the axle is a coil-over, the mounting angle of the coil-over also has to be fac-

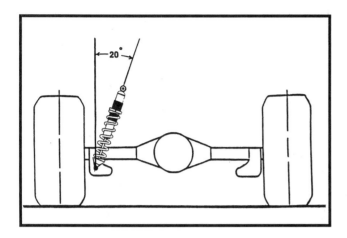

tored in to find the true wheel rate. Use an angle finder to determine the number of degrees from vertical that the coil=-over is mounted. Then find the cosine of that angle in a Trig Table, square the cosine, then multiply that times the spring rate times the motion ratio.

As an example, the spring is 250 lbs/inch, the motion ratio is .688, and the coil-over is mounted at 20 degrees from vertical. The cosine for 20 degrees is .94. So:

$$WR = (MR)^2 \times SR \times (\cos)^2$$
$$WR = (.688)^2 \times 250 \times (.94)^2$$
$$WR = .473 \times 250 \times .8836$$
$$WR = 105$$

A Trig Table can be found in the Front Suspension chapter of this book.

Cambered Rear Ends

Cambered rear ends allow wheels to rotate in a plane that is not perpendicular to the rear end housing. This helps get the tires closer to being flat on the track surface during cornering when tire stagger is present.

Rear end cambering is essentially most effective with radial tires. That is because radial tire sidewalls are so stiff.

Bias ply tire sidewalls are much more flexible, but cambering can still have a small positive effect on them. Bias ply tires develop maximum cornering forces when cambered at –1°. This means stagger coupled with negative camber can create maximum cornering characteristics.

A maximum of –.5° camber at the right rear and +.5° camber at the left rear can be used with bias ply tires to help enhance cornering. The use of stagger puts some negative camber in the right rear and

A rear end housing snout machined for 1.5-degrees of camber. Note how it is offset.

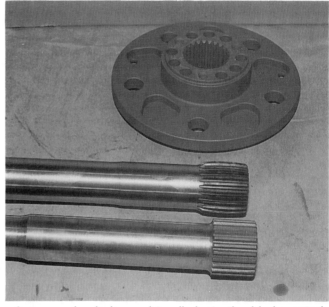

A crowned axle is used to eliminate the bind created between the axle spline and drive flange when a cambered snout is used. A special heat treated and coated drive flange is required.

positive camber in the left rear. In general (depending on sidewall construction and tire pressure), 1-inch of stagger creates .25° of camber. So 3 inches of stagger will produce .5° to .75° of negative camber at the right rear.

With radial tires that have much stiffer sidewalls, usually the right rear camber is set at –1.5°, and the left rear is set at +1.5°. Also, the right rear is set toed-in .03125-inch in order to keep the radial tire

The Super Max quick change from Speedway Engineering. Note the oil line kit which pumps lubricant to the center of the pinion gear and bearing.

sidewall loaded to ease the transition to cornering. When using this much rear end camber, flexible upper links cannot be used in the suspension. This is because the flexible link allows the rear end to wrap-up under acceleration. When it does this, a negatively cambered hub rotates with the housing and becomes a toed-out hub instead.

In many racing associations, the use of cambered rear ends is prohibited. Check your rules before ordering a rear end housing.

The Rear End

There are two different types of rear ends to use – a quick change, or a stock passenger car type of rear end. For pure competition, the quick change is the only type to consider. However, many racing classes are limited by rules to the stock type of rear end.

A quick change rear end allows the racer to more easily match the gear ratios he needs to fine-tune to a particular application It is especially helpful when the racer runs more than one track.

The Ford 9-inch can be a more efficient rear end if the gearing is available for a particular application. This is because the Ford 9-inch weighs less than most quick changes, and it takes less power to turn than a quick change.

But if you are looking for power efficiency and lighter weight with the benefits of a quick change, consider the Super Max quick change.

Note how much smaller the Super Max ring and pinion (top) is compared to a regular quick change ring and pinion (bottom).

The Super Max Quick Change

Speedway Engineering re-engineered the traditional quick change several years ago with the goal of reducing overall weight and rotating weight while keeping it as durable as ever. Their very successful product is called the Super Max quick change.

The Super Max weighs only 134 pounds ready to race as compared to 175 (or more) pounds for a typical quick change. In addition, a Super Max Detroit Locker® with ring gear attached weighs only 16 pounds, whereas a Ford 9-inch ring gear by itself weighs 15 to 20 pounds (depending on the ratio). This is a considerable reduction in rotating weight. The Super Max uses an aluminum Detroit Locker® carrier, or an aluminum spool, to keep the rotating weight lower.

The Super Max ring gear has a smaller diameter than a regular quick change, and it is of a semi-hypoid design. This type of gear design is required for the smaller diameter in order to get efficient gear contact. The hypoid type of gear design can be less efficient because of an excessive amount of gear contact. But the Super Max gear has a very efficient design. It has less gear contact than a Ford 9-inch ring gear, so it requires less horsepower to turn than the 9-inch gear.

The Super Max was designed to handle 550 horsepower or less. It is suitable for Southwest Tour, Northwest Tour, Busch North, and similar series, and almost all Saturday night sportsman and late model cars. The less horsepower a car has, the more beneficial the Super Max's lower rotating weight is. It will handle slightly more than 550 horsepower with regular servicing and attention.

The pinion gear in the Super Max sets above the centerline of the ring gear, so it doesn't get a lot of lubrication. The rear end comes with an oil line kit (which hooks up to your pump) that pumps lubricant to the center of the pinion gear and the pinion bearing. Any car that runs 100-lap races requires a rear end lubricant pump to help the components live, regardless of what type of rear end is being used.

Differentials

A differential, in general, has two primary functions during cornering – to differentiate wheel speed from left to right while cornering, and to distribute torque to the wheels for proper traction.

Speed differentiation occurs when a car is in a turning or cornering mode. The outside wheel must increase its speed relative to the center carrier, and the inside wheel must decrease its speed a proportionate amount. This is because the outside wheel travels a greater distance than the center carrier while the inside wheel slows down.

Torque distribution is important because that function dictates how power is delivered to the left and right wheels based on the amount of traction each has. What is important for racing applications is that torque is diverted away from the wheel that wants to spin, and it is delivered to the wheel that has more traction.

One of the major reasons for using a differential in place of a spool is the ability to use less stagger. If a spool is used in the rear end, more stagger must be used to turn the car. The amount of stagger is tuned to adjust for rear end induced looseness or push. However, once a differential is installed, the left rear is disconnected from the right rear during left hand cornering. The forward drive and drag caused by the spool at the left rear is not present, so stagger can be reduced. When less tire stagger is used, there is less drag on the rear tires down the straightaways. This allows the rear tires to run cooler and lessens rear tire wear. This is very important in longer races as the track surface gets slicker and tires wear.

There are a variety of differential types – open, fully locked, locker or ratchet, torque sensing or torque bias, and torque reactive.

The Open Differential

The open differential is the typical passenger car differential without any type of limited slip capabilities. Its sole purpose is to differentiate, which means it divides the speed between the inside and outside wheels during cornering. It offers no torque distribution. That means the torque gets applied to the wheel which has the least resistance or traction. Under cornering and acceleration, the open differential delivers all the torque to the least loaded wheel, and the result is wheel spin. This prevents power from being delivered to the wheel which has the most traction.

The Locked Rear End

The locked rear end uses a locking spool or mini spool to tie both rear axles solidly together. A locking spool replaces the spider gears in a differential. With the locking spool installed, there is no differentiating action or torque distribution.

The locking spool is ideal for straight line acceleration. But, it presents some problems during cornering. When a racer decelerates and turns into a corner, both the inside and outside rear wheels want to continue turning at the same speed. Because they are tied together, nothing else can occur. The problem is that the rear wheels will resist changing from their straight ahead direction. To help the rear end overcome this resistance to change direction, tire stagger is used. The problem encountered with the use of excessive tire stagger (which overcomes directional change resistance) is higher right rear tire temperature and excessive tire wear. The chosen stagger also commits the driver to one particular groove.

The reasons that the locking spool has such wide spread use in stock car racing, however, is that it is simple, it is durable, and it is lightweight.

The Locker Differential

The locker differential (also called a ratchet) provides differentiating. A locker differential uses spring-loaded ratchets which lock up solidly under power, and unlock the outside wheel when the

An important modification of the Detroit Locker® for racing use is to remove the holdout ring on each side (see arrow).

The operating parts of a Detroit Locker®: the central driver (upper right), the 2 driven clutch assemblies (to the left of the central driver), springs (left), and side gear (far right).

power is off. The ratchets operate independently side to side.

The Detroit Locker® is either fully locked up or has one wheel unlocked and freewheeling. There is no in-between. Under deceleration and corner entry, the outside wheel is disengaged. Under acceleration – even with the throttle lightly feathered – both wheels are locked up and drive equally. The proper amount of stagger is critical with a Detroit Locker® because under acceleration the two rear wheels are locked together just like with a spool.

The Detroit Locker® differentiating system uses dog gear teeth or ratchets. When under full power, the dog teeth of the two driven clutch assemblies are fully locked together with the dog teeth of the central driver. When one wheel begins to roll faster than the other (such as cornering), the outside driven clutch is disengaged from the central driver by means of cam ramps on a ring inside the central driver. As soon as power is applied, the dog teeth lock back together and the speed of both rear wheels is equalized. That is why stagger is an important element when a Detroit Locker® is used.

The disengagement of the outside wheel at corner entry creates a free wheeling effect which allows a smooth corner entry as opposed to the drag created on the outside tire with a spool. This allows the driver to adjust his line into and through a corner.

A driver has to be smooth in his style when using a Detroit Locker®. At the middle of a turn, if the driver gets on and off the gas, it will alternate be-

tween free wheeling and full lockup. This will loosen up the rear of the car and may cause it to spin.

If a driver uses heavy trail braking entering a turn, the outside ratchet may remain locked up, sensing power application instead of power decrease. In this case, the differentiating will not take place, and the car will push. This also gives the driver an option to change handling characteristics if a car is loose at turn entry.

An important modification of the Detroit Locker® for racing use is to remove the holdout ring on each side. This assures quicker lockup of the cogs on the driven assemblies. With the holdout ring installed, the driven clutch has to make a full revolution before it engages with the central driver. With the holdout ring removed, the two assemblies lock together at the next cog encountered. This means almost instantaneous lockup when the driver gets on the throttle. There is no waiting for engagement when the driver hits the gas.

The rate of the springs which bear against the driven clutches is critical. The rate of the springs does not affect how quickly the locker reacts. The only function of the springs is to keep the driven clutch assemblies properly mated together with the central driver so that they do not unlock under partial throttle conditions or when bumps are encountered. However, the Locker will not function correctly with different spring rates side to side. The spring rates much be identical on both sides, when loaded at the proper operating height. These springs (in extra duty stiffness) are available paired together with identical rates checked on a digital spring rate checker from Tex Racing Enterprises.

The Diamond Trak from Quick Change Exchange is a torque sensing differential.

The Gold Trak differential from Dan Press Industries.

Torque Sensing Differentials

A torque sensing (or torque bias) differential simultaneously provides each rear wheel with the RPM difference required to compensate for turning radius and the torque distribution required at each wheel for proper traction. It senses and redistributes torque to allow for a variation in traction at the two rear wheels.

The most important characteristic of a torque sensing differential is that it will increase torque to the wheel with traction. This is opposed to an open differential that transfers drive torque to the wheel that is turning the fastest. The torque sensing differential senses wheel slip and delivers torque to the wheel with the greatest amount of traction. It does this by using spiral pinion gears to connect the axles to each other, directing the most force to the axle with the most resistance.

The advantage of the torque sensing differential is that as a car enters a corner, neither rear wheel is completely unlocked, as is the case with a locker. There will still be some torque loading of the outside wheel. This keeps the car tighter going into a turn, which is a more stable feeling for some drivers. This also means less stagger is required to turn the car as compared to a spool. And this means less right rear tire heat and wear.

Torque sensing differentials use spiral cut gears which create a self-locking effect when under load. When loaded, the spiral cut of the gears makes them thrust hard against each other and the carrier housing, creating friction.

The Gold Trak Differential

The Gold Trak differential made by Dan Press Industries (DPI) is a torque sensing unit which uses an arrangement of helical pinion gears to create differentiating while both rear wheels receive torque in relationship to their traction. The unit uses pairs of these gears which must stay in balance with each other. If the gears sense an imbalance in torque between the two wheels, the pinion gears redirect the torque bias toward the wheel with the greater traction.

The AFCO E.T.S. Differential

The AFCO E.T.S. is a torque sensing differential that uses worm gears set perpendicular to each other. There are six evenly spaced side gears which have worm gear interiors and spur gear ends. The worm gear portions mate with the worm-type axle gears at a 90-degree angle.

As a car corners, the outside wheel speed increases. When this happens the axle worm gear turns the mated side gears. The inside axle is slowed, thus slowing the inside axle worm gear in proportion to the outside. This slows the side gears mated to the inside axle gear. The spur gears on the ends of the side gears mate with each other and balance the two axle gears to each other to provide the differential action.

The AFCO E.T.S. torque sensing differential.

The Variloc system installed in a Ford 9-inch application. The ramp assembly is on the right, the steel clutches on the left.

applied to both wheels in proportion to the traction available at each wheel. The largest split of torque that it will allow is 70 percent to the wheel with more traction and 30 percent to the other. It will never shift 100 percent of the torque to one wheel.

The Variloc Differential

The Variloc uses torque to produce a clamping action on an alloy steel clutch pack. The amount of locking produced is directly dependent on the amount of torque load and the internal components used. The Variloc is a "torque reactive" differential

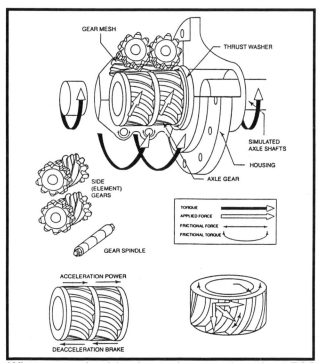

When one wheel begins to lose traction, the AFCO E.T.S. directs the torque to the wheel with the most avialable traction.

When one wheel attempts to spin, the corresponding axle gear tries to increase its speed. As it does this, it tries to turn the mated side gears faster. Since the side gears are mated to each other through the spur gears, and the wheel with traction is driving its mated side gears, the side gears on the loose wheel cannot turn the side gears on the wheel with traction. Since torque is applied in even proportions, it is

Variloc ramps cut to different angles or lockup rates. (L to R) 45x45, 45x70 and 60x60. Higher numbers denote lower lockup rates.

because of the way its locking ramps clamp clutches in reaction to torque loading.

Lockup is produced as a reaction to torque loading. The ramp and spider gears act as wedges driven by the torque load. When power is applied, the spider gear tries to climb up the angular surface of the ramp, but because it is constrained by another surface, it forces the ramp to thrust sideways and clamp down on the clutches. The amount of lockup produced depends on the angle of the ramp surfaces and the amount of torque applied. Lockup is not affected by clutch wear or ramp/spider gear interface. As clearances increase, the spider gear just climbs further up the ramp surface.

By varying the angles cut into the ramps where they contact the spider gears, different amounts of locking action can be achieved.

The Driveshaft

The standard driveshaft for most Saturday night stock cars is a 3-inch O.D., .065-inch wall D.O.M. 1020 steel tube. Most track rules — including the NASCAR Winston Racing Series – requires this type of driveshaft. The material must have close tolerances in wall thickness and O.D. runout to help in balancing the shaft.

Driveshafts can easily turn 6,000 to 7,000 RPMs, and more. So they have to be balanced in order to tolerate those speeds. An out-of-balance shaft will flex in the center like a bow, and snap when the stresses get too extreme. Out-of-balance driveshafts also have a tendency to tear up transmissions. All driveshafts should be precision balanced before being used in competition.

High quality universal joints are a must in order to assure a fully balanced driveshaft. The higher quality is seen in improved material alloys and machining tolerances of the part. A good competition u-joint is Spicer's 1310 series.

If you have a vibration in your race car, suspect the driveshaft first. Take it out and have the balance checked. The vibration harmonic frequencies can quickly ruin u-joints.

Aluminum driveshafts – when allowed by track rules – provide a reduction in overall weight and rotating weight. A driveshaft that is lighter requires less engine power to rotate it.

The most common aluminum tubing used is a 3-inch O.D., .25-inch thick wall. This type of alumi-

Driveshaft safety loops should be considered mandatory, even if track rules don't require them.

num driveshaft will save about 8 pounds over a comparable steel shaft.

Driveshaft safety loops should be considered mandatory, even if track rules don't require them (but most do). They are important safety items in case of a driveshaft or u-joint failure. Most associations require two loops – one placed six inches behind the front of the driveshaft, and one six inches forward of the rear of the driveshaft. IMCA modified rules require the loops to be fabricated from .25-inch thick, 2-inch wide steel material. This is a very excellent choice. Check with the rules for your track.

Another important safety consideration is to paint the driveshaft white. This makes it more visible should a driveshaft come out of the car and land on the race track.

Inspecting The Driveshaft

Regular inspection of the driveshaft and u-joints is very important to assure good driveshaft life and safety. When the car is first assembled, move the rear suspension through a full maximum range of travel to be sure the driveshaft does not contact anything.

Each week, check the driveshaft for any gouges, dents, or dings. These can weaken the material and affect balance as the shaft spins. If you find anything, replace the shaft or have the manufacturer inspect it. Also check the u-joints for play on a regular basis.

Pinion Angle

The pinion angle is the angle of the pinion gear in relationship to a line drawn straight ahead at the

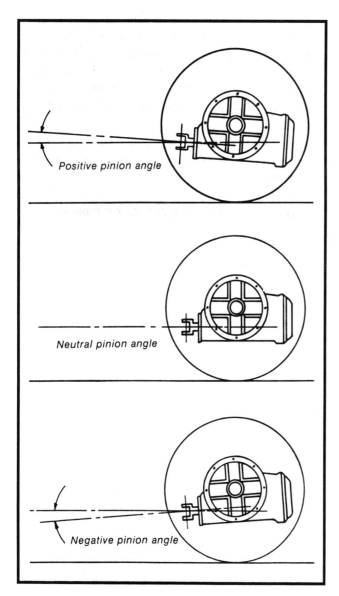

Positive pinion angle

Neutral pinion angle

Negative pinion angle

Rear End Maintenance

Taking good care of the rear end lubrication is one of the most important things you can do. Metal particles, dirt and grit can accumulate in the gear lube. When they do, they can cause gear abrasion and surface damage. Left unchecked, continued abrasion will wear away gear surfaces and cause excessive backlash. This will lead to gear failures. Metal particles floating in the lube can also destroy bearings. Rear end lube that has been overheated will not provide the required protection for gears and bearings.

The easy cure is to change rear end gear lube on a regular basis. Gear lube should be changed after every five hours of racing operation. If you use a torque sensing type of differential, the change interval should be shorter because they operate much hotter than spool or locker equipped rear ends.

If you use lubricant cooling in the rear end, be sure to use a filter between the rear end and the cooler. Install an inline 75-micron screen filter. Filtering keeps microscopic metal particles and grit out of the lubricant, protecting gears and bearings. A manufacturer of rear ends we talked to reported that when he gets units back for rebuilding, he can tell right away if a filter had been used with the cooling system.

Other maintenance items for quick change rear ends include:

1. Check the tightness of the pinion nut. If it has loosened up, the preload is gone on the bearings.

2. Check to be sure the rear cover bearings are free.

3. If you should break a quick change gear, replace the bearings and the cover.

4. Once a month, pull a side bell and check for ring gear pitting.

5. Have all gears and shafts Magnafluxed once a year. These parts wear out and should be replaced before they fail. Gear and shaft failures during races can cause enough damage to destroy the entire rear end. Check the lower shaft and pinion shaft for straightness.

6. Rear ends are normally rebuilt once a season. However, if you suspect accelerated wear because of metal residue in the gear lube, rebuild it. It's cheaper now. If you run a really tough schedule, rebuild more often.

7. Hub bearings should be checked every 5 races or so.

8. Check the lube level before every race.

universal joint coupling. A negative pinion angle has the front of the pinion gear pointing downward.

If there is excessive pinion angle, the driveshaft yokes can bind and cause damage to all driveline components. The proper amount of pinion angle depends on the amount of rear suspension travel that will be encountered, the type of track surface and the type of rear suspension being used. If a car has a 3-link rear suspension with a solid upper link, the pinion angle can be 3 degrees down. If the upper link is a rubber biscuit type of link, it will need 4 degrees down. With a spring-loaded upper link or with leaf springs, 5 to 6 degrees down will be required.

9. If you experience quick change spur gear failure or excessive wear, one of the following is probably the cause:

A. Wrong ratio for track conditions (teeth too fine).

B. Improper or bad gear lube.

C. Not replacing gear spacer when changing gears.

D. Crashing and bumping, especially with power on.

E. Overheating the rear end.

F. Mixed gear sets (gears are in matched sets).

G. Worn or loose rear cover bearings.

H. Bent pinion shaft or lower shaft.

I. Loose rear cover.

10. Remember to put quick change gears on the correct shaft. For ratios lower than the ring and pinion ratios (higher numerically) the big gear goes on top (and vice versa).

Rear End Lubricants

What is the best lubricant to use for racing? The most important answer is that any 90-weight gear lube rated for extra duty will work IF the rear end is set up correctly.

A 90-weight gear lubricant is a minimum. If a lower weight, thinner viscosity lube is used, there is a danger that it will not properly cling to the gears.

When a new or rebuilt rear end assembly is first used, take the time to break it in properly. Run some easy laps to get some heat into the gears and lubricant before putting any heavy loading on the gears. Regardless of how careful the rebuilder was, there will always be some dirt and metal particles in the lube after it has been run for the first time. Don't run any more than practice, qualifying and a heat race with the dirty lube. Change it before running a main event.

Many rear end building professionals feel that the initial break-in should be done with a petroleum based grease, even if you normally use a synthetic lubricant.

Synthetic Gear Lubricants

Synthetic lubricants are more costly than petroleum lubricants, but the synthetics – such as Mobil 1 – offer a lot of advantages that can justify the cost. The primary advantages are much better heat dissipation and better flow and viscosity characteristics. That means that when the synthetic lube is cold, it flows very fluidly. It is able to lubricate better than thicker petroleum lubes. And with better viscosity, the synthetic lube's thickness (viscosity) does not change measurably as temperatures increase dramatically.

The synthetics also do a better job at transmitting lubricant heat to the case and/or cooler, which helps dissipate heat much more efficiently. The end result is that racers report rear end temperatures generally run about 20 degrees cooler with synthetics.

Chapter

4

Shock Absorbers

There are two areas of race car technology that tend to leave a number of questions unanswered – tires and shock absorbers. These are "mystery" items to most racers, and yet they are the true keys to tuning the suspension to the race track.

Racers who understand shocks, and how to use them to fine tune their chassis, will have a distinct advantage. Racers are always looking for an edge, and they can find it through tuning with shocks.

The Purpose of Shocks

Shock absorbers – what are they? First of all, they are not **shock** absorbers! (The United States is the only place that calls them that. The rest of the educated world calls them "dampers.") **Springs** absorb shock. Shocks really **dampen** the kinetic energy that is stored in the springs. The energy is created by vertical movements of the wheels relative to the chassis. The kinetic (moving) energy is converted by the shock absorber to heat energy by creating resistance to movement within the shock.

Shocks are intended to control the vertical oscillations of the suspension caused by bump and dynamic load transfer. Without shock absorbers, when a bump or load transfer is encountered, the vehicle would continue to oscillate (bounce) at a natural frequency determined by the ratio of sprung to unsprung weight at each wheel. This oscillation would cause the car to roll, pitch and yaw in a variety of ways, none of which are comfortable or fast. To eliminate this leap frog action, shock absorbers are used.

How Shocks Work

A shock absorber is a velocity sensitive hydraulic damping device using a piston to move through a fluid-filled chamber. A shock absorber consists of a tube or shell, a shaft, and a piston connected to the shaft. The piston moves through oil (shock fluid) inside the shock tube according to input movements from the shaft. Suspension movement creates the force that is input into the shock absorber. The shock

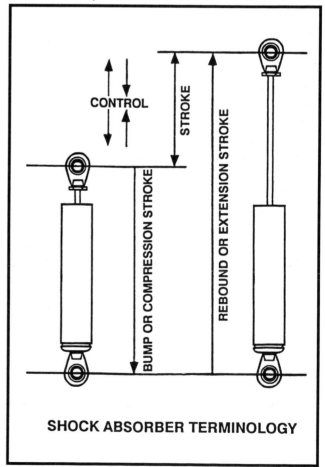

SHOCK ABSORBER TERMINOLOGY

absorber responds with resistance to piston and shaft movement by forcing the piston through fluid. The shock fluid is not compressible, so the piston must have holes in it to allow the fluid to flow from one side of the piston to the other as the piston is pushed or pulled through the body, or else hydraulic lock would result.

The shock fluid is forced to flow through valves and orifices in the piston. The oil causes a resistance through the piston as it moves up and down. The valves in the piston control the rate of resistance, and thus the rate of suspension damping. Shock absorbers are rated by the pounds of force required to move the shaft at a certain speed.

The rate of resistance through the valves can be tailored by incorporating holes of different sizes and valves with different spring rates. These create different stages of valving control. The staged valving regulates the damping control as the piston moves at various speeds.

As the piston passes through the shock fluid, the fluid is forced to pass through a series of orifices (small metered holes) and valves. Some of the orifices are open, such as the low speed bleed control, which permit the fluid to pass through during any type of piston movement. Others are blow-off type valves which open only when the pressure reaches a certain level. A blow-off is a spring-loaded pressure relief valve. It opens when sufficient pressure is built up to overcome the valve's spring pressure. Blow-off valves are also tailored to their particular application by the diameter size of the orifice they control Once the valve blows off, the amount of shock fluid flowing through the orifice depends on the diameter of the orifice. A smaller diameter produces a stiffer resistance at a given shaft velocity.

The stiffness of the shock is controlled by the size of the orifices and the stiffness of the valve springs (which time the valve opening and blow off). These design elements can be combined to create staged valving. This staged valving can create a different resistance to movement at lower shaft speeds, mid range shaft speeds, and high speed shaft movements. This way the shock can react one way during low speed body roll and another way when a wheel hits a high speed bump. This allows the shock's resistance to movement at various shaft velocities to match the needs of the race car.

It is important to understand how shock absorber shaft speed works in relationship to the amount of damping resistance the shock provides. The rate of acceleration and deceleration that the piston goes through is what creates different damping rates. For example, if a shock moves from 0 to 10 inches per second, the speed at mid stroke is 10 inches per second. But, at the beginning and end of that piston movement, the piston speed is 0. So, shock control depends on how quickly resistance is offered to a movement as the acceleration rate builds up. If the rate of force build-up was linear, that means the force would continue to build at the same rate, so resistance at faster piston velocities would be extremely stiff. For that reason, blow-off valves are employed which develop blow-off curves.

To have good damping control at the 10 inches per second speed, it is important for the shock to accelerate the pressure build-up quickly to that speed, and then blow off that pressure without a lot more pressure build-up past that control point. When this is accomplished, it yields good speed sensitivity.

Shock Valving Characteristics

When a piston with holes in it is pushed through shock fluid, a resistance force is generated. This is the basic premise of shock absorbers. With fixed openings in the piston, the resistance force will quadruple when the shaft speed doubles. So what might be an acceptable force at low speed would be unacceptably stiff at faster shaft speeds. So methods had to be developed to release or blow off the pressure after it reached certain levels. This is the reason for developing shock absorber control curves.

Shock valving characteristics can be linear, digressive or progressive (see graphs next page). Progressive means that the damping force builds up slower than the piston velocity. With linear valving, the damping force builds up at the same rate as the piston velocity. Digressive valving allows the damping force to build up quicker than the piston velocity.

The digressive rate shock is the ideal valving for a racing shock. It has good damping control (quick force build-up) in the lower range (0 to 10 inches/second of shaft movement) and not excessively high damping in the higher ranges (12 to 25 inches/second of movement). See the adjacent

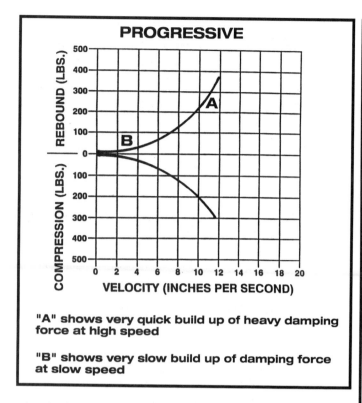

PROGRESSIVE

"A" shows very quick build up of heavy damping force at high speed

"B" shows very slow build up of damping force at slow speed

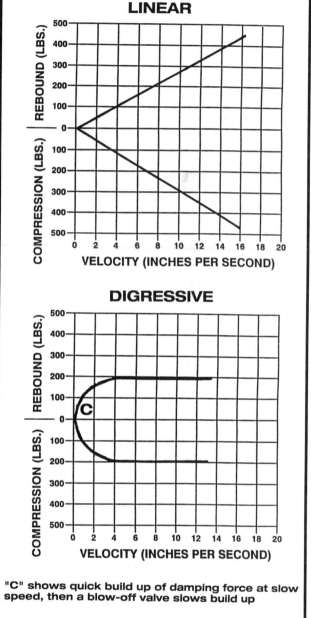

LINEAR

DIGRESSIVE

"C" shows quick build up of damping force at slow speed, then a blow-off valve slows build up

shock dyno chart for a representation of desirable damping control curves.

Shocks can be valved so that they have a combination of these characteristics in both compression and rebound. For example, under compression, the initial piston movement may build the damping force very quickly (progressive), but the force build

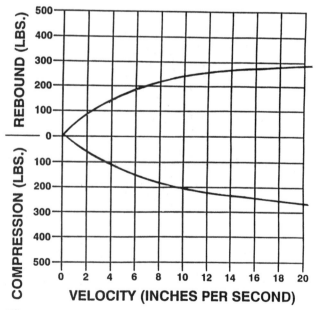

This is a representation of what desirable damping control curves should look like.

up has to trail off to linear (or digressive) or else the shock will quickly become too stiff as velocity increases. The build-up rates are called acceleration ramps. Different acceleration ramps can be at work at the same time in the shock, having one controlling 0 to 5 inches per second speed, and another controlling 5 to 11 inches per second speed, and each ramp accelerating at different rates.

To change the damping characteristics between the two acceleration ramps, blow-off valves are used. A blow-off valve is a pressure relief valve controlled by the spring rate in it. It controls how far

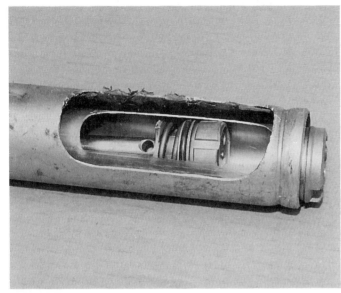

This cutaway of a double tube shock shows the outer tube, the inner cylinder and the piston and shaft inside the inner cylinder. The bottom end of the shock has been cut off and the foot valve on the inner cylinder protrudes outward. The space between the outer shock tube and the inner cylinder is the reservoir area. Visible at the top between the two is the gas-filled plastic bag. It appears frayed because we cut it during sectioning.

the valve lifts off the seat. It is much like the pressure relief valve in an engine oil pump.

These tailored blow off curves are unique to double tube shock absorbers because they employ blow-off valves. High pressure single tube shocks utilize a blow-off plate system, which primarily produces only straight line damping force.

Single Tube and Double Tube Shocks

There are two different type of shock absorbers commonly used in race cars – the high pressure gas shock (also called the monotube or single tube shock), and the low gas pressure shock (also called twin tube or double tube shock). Shocks such as the Koni, Bilstein and Penske are the single tube type. Shocks such as AFCO, Carrera and PRO are the double tube type.

Double Tube Shocks

Double tube shocks utilize two tubes – an inner cylinder where the piston operates against the shock fluid, and the outer shock body. The space between the two cylinders is utilized as a reservoir.

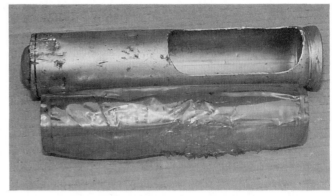

Below this shock outer tube is the gas-filled plastic bag that resides in the reservoir between the two tubes.

A shock cannot be fully filled with fluid. This is because the fluid has to have some place to go as the shaft moves inward in compression which subtracts from area inside of the cylinder. The fluid is non-compressible, so it has to have a chamber to travel to. In the double tube shock, that space is between the inner tube and outer tube.

If that space was filled with just plain air, a big problem would occur: aeration. Aeration occurs when air mixes with the oil and forms bubbles in it. This now makes the shock fluid compressible and

The foot valve at the bottom of the inner cylinder of a double tube shock contains two spring loaded valves. The one on the left is a blow-off valve that controls compression damping. The one on the right is a directional control valve – it pulls open under rebound to fill the cavity under the piston. The milled slots function as low speed bleeds to allow fluid transfer from and to the reservoir.

The piston and shaft assemby from a double tube shock. The heavy spring is a compression bypass spring valve. The holes in the piston are outlets for bleed valves and high speed jets. The milled slots in the piston control low speed bleed.

The piston and shaft assembly from a single tube shock. Under the nut on the end of the shaft is the shim stack that controls the compression damping. Under the shims are the orifices that control fluid passage. The shim stack for rebound valving is on the bottom side of the piston.

changes its viscosity, and thus lowers the resistance of the flow through the valves and orifices. This takes away most of the control characteristics of the shock.

To prevent oil aeration, the reservoir space between the inner and outer tubes is filled with a gas-filled plastic bag. The bag contains nitrogen, which is compressible. The gas bag allows for fluid volume changes as the shaft moves in compression and rebound. The gas bag volume changes during shaft movement. The plastic bag totally contains the nitrogen and prevents it from mixing with the shock fluid.

As a double tube shock moves into compression, the shaft acts against the piston which pushes fluid out of the shock body. This fluid flows through the base or foot valve of the shock as it is displaced out

of the inner tube and flows between the two tubes, compressing the gas-filled bag. The foot valve has orifices and valves which control the compression characteristics of the shock, and it also offers additional stages of valving control for the shock.

Single Tube Shocks

Single tube shocks have a single chamber, with the shock fluid separated from a charge of highly pressurized nitrogen by a floating piston.

Just like the double tube shock, the single tube shock has to contain a compressible gas in order to compensate for fluid displacement during shock compression. This gas is contained in the upper part

A cutaway of a single tube shock. At left is the shaft with the piston. On either side of the piston is a stack of spring steel shims. On the right of the sectioned tube is the floating piston. To the right of the floating piston is the chamber that is filled with pressurized nitrogen.

A Koni single tube shock with the shim stacks disassembled. The fluid passage orifices are visible in the piston. The tang wheel (lower left) engages the rebound shim stack to change rebound damping characteristics. The shock shaft is compressed and twisted to change the valving.

of the shock tube, under high pressure, and is separated from the shock fluid by the floating piston.

There are metering orifices in the piston that restrict fluid passage during compression and rebound. Valves in the form of a stack of spring steel shims are attached to the piston. Fluid flowing through the metering orifices must deflect the shim stack, which uncovers other orifices in the piston to alter the pressure build-up. A shim stack resides on either side of the piston, so one stack is deflected during compression (controlling compression damping force), and the other one is deflected during rebound, which controls only the rebound damping. The damping characteristics are determined by a given amount of resistance at a given piston velocity. These characteristics are tailored by the shape, diameter and thickness of the steel shims. Also controlling the damping characteristics are low speed bypass ports in the piston.

Under compression, the piston shaft enters the shock body, adding to the volume inside the chamber. The shock fluid is not compressible, so the total volume in the chamber increases and pushes against the floating piston. This compresses the pressurized nitrogen above the floating piston. Under rebound, this pressurized gas will expand to take up the volume of the retracting shaft in the chamber.

A single tube shock is vulnerable to damage caused by contact from debris or suspension components. Any type of dent in the shock body will cause the shock not to function because the piston will be pinched by the dented cylinder wall. And,

because of the addition of the gas chamber to the shock body, a single tube shock is longer, and thus slightly heavier than a double tube shock.

Damping Stiffness Codes

Manufacturers of double tube shock absorbers place a number on them to indicate the relative stiffness of the shock's damping force. This is called the valve code. The numbers range from 2 to 8, with a 2 being a very soft shock and an 8 being a very stiff shock.

In most manufacturers' numbering system, a single digit valve code (such as a 5) designates a 50/50 ratio shock. This means – theoretically – that the shock absorber is valved to offer the same amount of resistance in rebound as it does in compression. In reality, some manufacturers valve their shocks to supply heavier resistance in rebound. For example, a typical PRO 5 series shock with a piston speed of 6 inches per second will show 185 pounds of damping force in rebound, and 140 pounds of damping force in compression.

When the valve code is designated as a double digit, such as 57, it is called a split valve shock. The split valve shock contains the valve code characteristics of the first number in compression, and the second number in rebound. For example, a 57 shock reacts like a 5 shock in compression, but like a 7 shock in rebound. This type of a shock allows you to tailor the handling and reaction characteristics at a particular corner of the car during weight transfer and body roll.

The PRO rebound adjustable shock has the rebound control valve located in a separate body on top of the fluid reservoir. The rebound valving stiffness is chosen by rotating the adjuster slot.

Adjustable Double Tube Shocks

Several of the manufacturers of double tube shocks offer an externally adjustable shock absorber. Some are adjustable in rebound, some are adjustable in both rebound and compression. All use different methods to accomplish the adjustments.

Adjustable shocks provide the racer a cost savings because a wide range of split valve shocks are not required. One or two different shocks per corner are all that is required to fine tune the handling of the race car.

PRO Rebound Adjustable Shocks

There are four factors that comprise a shock valving code in a double tube shock: 1) Low speed bleed, which is the basic low speed control; 2) Spring rate of the spring that operates the bleed valve; 3) The seat pressure on the spring and valve; and 4) High speed jet, which controls the high speed valving.

Previous to the introduction of this shock, externally adjustable shocks had only adjusted the low speed bleed and the seat pressure on the spring. The PRO rebound adjustable shock changes all four

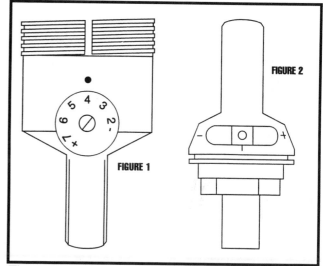

The compression valving in AFCO's double adjustable shock is changed with the knob on the bottom of the shock body (figure 1). The rebound is adjusted by twisting the adjuster ring in a window at the top fo the shock shaft (figure 2).

elements of the valving code. This is accomplished by locating the rebound control valve in a separate valve body on top of the fluid reservoir. This valve incorporates a one-directional flow valve, so fluid flows in for control, then drains back to the reservoir. This helps to recirculate the fluid and thus helps to cool it.

The PRO shock is totally adjustable for rebound control. One revolution of .084-inch can adjust the rebound from a 3 valve code to an 8 valve code. The rebound damping is infinitely adjustable to any valve code between 3 and 8, being able to stop at any point between the single digit numbers. This allows the racer to fine tune the rebound to valvings such as 4.5 or 5.75. Each of these valve code numbers produce a rebound force exactly the same as PRO's regular shocks.

The proper shock for an application is chosen by picking the compression damping desired. The adjustable shocks are available in 3, 4, 5 and 6 compression valve codes, in both smooth and threaded aluminum bodies.

AFCO Double Adjustable Shocks

AFCO's double adjustable shocks have independent adjusters that allow rebound or compression changes without affecting the other valving adjustment.

The adjusters on AFCO's double adjustable shocks.

The compression valving is adjusted with a knob located at the side of the bottom end of the shock. It adjusts from a 2 valve code to a 7 valve code. The compression can be set between the valve numbers for fractional valve settings, such as 4.5, 5.5, etc., without fear of the adjuster knob rotating during a race.

In both compression and rebound, each of the valve code numbers produce a damping force exactly the same as AFCO's regular shocks.

The rebound adjustment is located in a window at the top of the shock shaft. Holes in the adjuster rod represent the valve stiffness number. The rebound can be adjusted from a 2 valve code to a 7 valve code. The adjuster ring is rotated until the desired valve number is lined up with an indicator mark on the housing. The rebound can be set between the valve numbers for fractional valve settings, such as 4.5, 5.5, etc., without fear of the adjuster rod rotating during a race.

The double adjustable shocks are available in 7 and 9-inch stroke lengths in a threaded aluminum body.

Carrera Adjustable Shocks

Carrera's 81 series 2-inch OD threaded aluminum body shocks offer two different types of external adjustment – shock shaft adjustment and external reservoir adjustment.

The shock shaft adjustment is accomplished by a twisting movement of the shaft which moves internal valves, thus changing the flow characteristics through orifices. This changes both bump and rebound damping forces.

The shock's damping operation is changed by altering the rebound and compression ratio. In the "1" position, the shock is a 35/65 (rebound/ compression) ratio. At the "2" position, it is a 50/50 shock. In the "3" position it is a 70/30 ratio (rebound/ compression) shock.

The shock is adjusted by fully depressing the shaft, then rotating it. Turn the shaft clockwise until you feel the adjustment mechanism engage, then continue rotating until the mark on the shaft lines up with the appropriate mark on the shock body.

The remote adjustable 81 series shock has a twist knob located on the remote reservoir that can change compression damping between a 4, 5 and 6 valve code. The rebound is internally adjustable.

Shock Absorbers and Handling

Shock absorbers are extremely important to a car's handling. Shock absorbers control the rate of weight transfer during cornering, they control spring movement, and they control suspension movement over bumps and surface undulations. Being able to control the chassis with the proper shock absorbers is a key element to proper handling. Shocks can be used to help control handling problems or to induce desirable handling characteristics.

Shocks have no effect on the *amount* of weight that is transferred dynamically during braking, acceleration and cornering. They can, however, affect the rate of response in the pitch and roll axis. The amount of weight transferred is dependent on the center of gravity, roll axis, and roll rates. *How quickly* the weight is transferred is controlled by the shock absorbers. This may only be for an instant, but

Differences in the primary stage valving of shocks can be used to tune the handling response of a car as it transitions from straight line to a cornering mode.

If the car is loose at turn exit, a decrease in the rebound damping of the front shocks allows weight to transfer quicker from the front to the rear.

shocks play an important part in handling response during that instant. It is the low speed damping forces of a shock that most influence the chassis as it rolls or pitches (corner entry/exit, acceleration/deceleration).

By altering the primary valving in the shock (the low velocity stage), the transient handling response of a race car can be altered as it goes from a straight line to a cornering mode, and from braking to acceleration or vice versa.

The same four shock absorbers are not right for all conditions. There are a variety of shock ratings and compression/rebound valving splits available. Our job here is to help you understand what they do, how they work, and how they relate to what you want the chassis to do under different conditions on the race track.

The most important thing for your understanding of choosing the correct shock is how to analyze the race track, and how to relate that to the different functions of shock absorbers. When you understand the concept of how shock absorbers work and can relate them to different handling conditions, then you will know how to tune your chassis with shocks. This will put you far ahead of the guy who just looks at an application chart in a catalog and picks his shocks that way. Like we said before, the same four shocks are not right for all conditions.

Start with a suggested shock absorber setup (as shown in the **Chassis Setup** chapter) for your application, and then experiment to find the ideal setup. Try softer and stiffer shocks, and different combina-

tions. See if it feels better or worse to the driver. Driver feedback and lap times will be the ultimate decision maker for the correct shock setup. When you experiment, be sure to keep good notes.

Chassis Tuning With Shocks

The following are basic handling adjustment tips for tuning your chassis with shock absorbers. As you read them, think about the basic concept of what the shock is doing to the chassis. This way you will be able to make decisions at the track to cure handling problems that your car may encounter.

As you read through these shock absorber chassis tuning guidelines, try to picture in your mind how the car is reacting at each position on the track, and how the prescribed changes can affect the chassis. Visualize how the weight is transferring at each corner and at each position on the track, and how shocks can restrict or accelerate transfer at a particular track position. Once you understand how the weight transfer and transitional control process works, you will have a good understanding of how to fine-tune your chassis with shocks.

Loose At Turn Entry

1) Increase the right front compression. This keeps more weight at the left rear longer by slowing transfer to the right front.

2) Decrease the left rear rebound. This is most effective under braking because the right rear will have more influence on stopping the car.

If the car is tight at turn exit, an increase in the rebound damping of the front shocks will help to keep weight from transferring so quickly from the front to the rear under acceleration.

Tight At Turn Entry

1) Increase rebound on the left rear. This keeps weight on the left rear longer during corner entry, preventing a quick transfer onto the right front corner. The left rear is the most heavily loaded corner of the car, so it is the best corner to control with shock damping.

Loose At Mid Turn

1) Decrease the left rear rebound. This helps to continue loading the right front from turn entry.

2) Decrease the right rear compression. This allows weight transfer to roll over onto the right rear to stick that tire.

Tight At Mid Turn

1) Decrease rebound at the left front. That allows weight to transfer more quickly to the right rear.

Loose At Turn Exit

Decrease the rebound on front shocks. This allows weight to transfer quicker from the front to the rear. When combined with a softer compression shock at the left rear, weight transfers quickly from the front and settles at the left rear to tighten up the chassis.

Tight At Turn Exit

Increase the rebound on the front shocks. This keeps weight from transferring so quickly from the front to the rear under acceleration.

This double shear bracket for mounting the top of a shock absorber is very well designed. The right side spacer is welded to the bracket while the left spacer mounts through a large hole in the bracket. The shock mounting bolt clamps it all together. It is efficient because it eliminates the time required to line up the spacers every time a shock is changed.

Ideal Shock Absorber Mounting Locations

Front Shock Mounting Location

The more travel the shock absorber piston has per inch of wheel travel, the more effective the damping will be, regardless of the rate of damping. To achieve this, it is imperative that the front shock be located as close as possible to the lower ball joint. So many chassis builders hang the shock absorber on the side of the front A-arm. In order to gain wheel clearance with this type of mounting, the shock has to be positioned at least 50 percent of the A-arm distance back to the mounting point of the arm. This makes the shock absorber almost totally ineffective.

The proper place for mounting the front shock absorber with the Camaro front stub is just inside the lower ball joint on the lower A-arms. This is the closest point to the wheel that is possible with this frame. The side of the coil spring pocket on the frame will have to be trimmed (see photos) in order to accommodate the shock in this position. This will

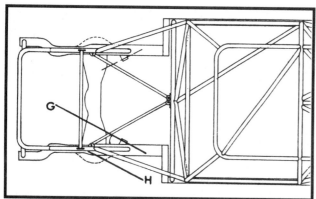

Shock absorbers must be mounted in the same place as the swing angle of the lower control arm. Here, "G" represents the swing plane of the lower control arm while "H" is the proper mounting plane for the shock absorber. If the planes are not parallel, suspension bind will occur.

It is imperative that the bottom of the front shock be located as close as possible to the lower ball joint.

not be detrimental to the structural integrity of the frame.

With the lower end of the shock absorber mounted as close as possible to the lower ball joint (on any type of suspension), position the top end so the inward mounting angle falls between 10 and 20 degrees from vertical. 15 degrees would be ideal. With the car setting at normal ride height, the shock

This shock mount is a tube which is installed through a hole drilled in the 1.75-inch OD vertical tube. The tube intersections are welded on both sides to create a very strong mount. The tube is tapped for a standard shock mounting bolt. This is a good mounting method for a shock, but it would not be appropriate for a coil-over. It is best to mount a coil-over in a straddle mount double shear bracket because it eliminates any bending loads on the mounting bolt.

absorber should be mounted at one half its total travel. The stroke of the shock should be positioned so the shock never reaches the end of its travel or bottoms out during the full amount of wheel travel.

Rear Shock Mounting Location

The rear shock absorbers should be mounted ideally at 8 inches below the center line of the rear axle housing. This is because it is easier to transfer weight front to rear when the top shock mount location is lower in relation to the CGH of the car, making it easier to compress the shock absorber (by greater leverage ratio between the CGH and the top shock mount height). The bottom mount should also be between 4 and 5.5 inches inward from the back face of the brake rotor.

The shock mounting angle should be 15 degrees from vertical. It should not vary more than a couple of degrees from this figure in either direction. The greater the inward mounting angle, the lower the motion ratio is on the shock. This means less shock piston travel for each inch of wheel movement, resulting in reduced shock absorber control of the wheel and chassis.

Check very carefully for any interference between the shock and suspension components.

With the car at proper ride height, the shock absorber should be at one half its total travel. The stroke of the shock should be long enough so that the shock never reaches the end of its travel or bottoms out during the full extent of wheel travel. If it does bottom out during compression on the race track, the spring rate will immediately jump to infinity (solid suspension) and you'll probably ruin the shock to boot. And, if it reaches the end of its travel on the inside wheels during extension, it will pull the wheels off the ground. Make sure there is plenty of shock travel.

Leverage Ratios On Shocks

It is common practice in some forms of racing to provide more than one upper (or lower) mounting position for a shock absorber. This is done to alter the motion ratio (by increasing the angle). That is fine for springs where we are simply dealing with mechanical leverage. However, shocks are hydraulic and the lowering of the motion ratio does not change the damping force of the shock absorber. It does, however, reduce the amount of piston travel and therefore the volume of fluid that is displaced. This in effect *reduces control*, not *damping* force. Maximum *control* is what is desired, therefore the shocks should be mounted as close as is reasonable to the wheel, and at an angle not to exceed 20 degrees from vertical in the front end, and 10 to 15 degrees in the rear end. If the damping force must be changed, change the shock and not the mounting position.

Rear shock absorbers should be mounted at 8 inches below the center line of the rear axle housing, and between 4 and 5.5 inches inward from the back face of the brake rotor.

Multiple lateral mounting positions that move a shock further away from the wheel are undesirable because the change in mounting position reduces damping control.

Dyno Testing Shocks

A shock dynamometer is a machine that measures a shock absorber's function. It is used to identify and evaluate the damping characteristics of the particular shock absorber being tested.

Shock dynos are highly sophisticated machines. They are attached to computers to aid in the evaluation of shock absorber functions. They can determine and chart an infinite number of characteristics. Damping forces (in pounds of force) are measured at several different piston speeds, from slow to fast. Charting how many pounds of force a shock generates at certain piston speeds (measured in inches per second) can demonstrate how the shock will affect the handling of a race car. Charting force versus piston speed will show curves that influence control characteristics at various piston velocities.

Shock dynos can provide a tremendous amount of information. But you have to be careful in charting the results from the variables in order to get useful, comparable data. The important variables are the stroke length and the stroke velocity (inches per second). The resulting resistance forces (in pounds) can be charted to show a comparison between different shocks.

The faster you stroke a shock, the stiffer it gets. Blow-off valves inside the shock tailor the transition ramps from one speed range to another (in double tube shocks). On a shock dyno graph, this should look like a smooth, flowing curve. The shock with the smooth flowing curve can handle more subtle changes in the track surface, as well as control body roll better.

The rate of acceleration and deceleration during shock piston travel is what creates different damping rates. The low speed and second stage damping ranges are the critical elements of a shock absorber. Different shock manufacturers have different philosophies about how these numbers should work and how the different damping ranges should phase together (called acceleration ramps). That is why different brands of shocks will make a chassis behave differently. Differences in valving design make a difference in chassis set-up.

For example, using shocks that have less low speed control requires the car to be set up with more left side weight and less cross weight. Using shocks that have more low speed rebound damping control allows less left side weight in the chassis set-up. This is because more low speed rebound control in the shock keeps more weight on the left side during corner entry.

The adjacent shock dyno charts illustrate these differences for you on the full range of AFCO, Carrera, PRO, Bilstein, and Koni shocks.

How We Dyno Tested Shocks

To create the shock dyno graphs illustrated in this chapter, we had an independent testing source dyno a full range of AFCO, Carrera, PRO, Bilstein and Koni shocks. All shocks tested were new shocks that came off the shelf of various dealers. No manufacturers were asked to contribute any products. All testing was done in a temperature-controlled environment so that all conditions were the same for all shock tests.

All shocks were tested on a Roehrig Engineering shock dyno. This is a very sophisticated dyno that collects 300 samples for 10 seconds. It shows the acceleration curve exactly the same as the shock sees it on a chassis. It measures resistance while accelerating to a peak number.

When dynoing shocks, to get a true reading of the shock's damping capabilities, each range of movement has to be measured. That means testing the shocks at each increment of speed, such as 1"/sec., 2"/sec., 4"/sec., 10,"/sec., etc., all the way up to 25"/sec. Many dynos are not equipped to handle that type of testing. They will typically test at just 3"/sec., 6"/sec., 12"/sec., and 25"/sec., then connect the dots of the peak readings at those speeds on the graph. But that doesn't tell the whole story. In between these testing points, the curve might flatten out or even dip. If the shock is not tested at all speeds through the entire range, the true ability of the shock to offer damping control is not known.

With the Roehrig shock dyno, readings are made as the machine accelerates the shaft velocity. This gives you the opportunity to see how the shock reacts as it accelerates slowly versus accelerating very quickly. As you can see from the adjoining shock graphs, the resistance force profile is different on every shock when it is accelerated at 10"/sec. and then at 25"/sec.

Each shock that we dynoed was accelerated to 10"/sec. and 25"/sec. The 10"/sec. velocity graph is shown on the left, and the 25"/sec. velocity graph is shown on the right for each of the shock brands tested. Careful observation of the graphs shows how each shock reacts differently at the two different accelerations. Look at the 5"/sec. and 7"/sec. force numbers on each graph and see the difference of what happens when the shock is accelerating at 10"/sec. versus 25"/sec.

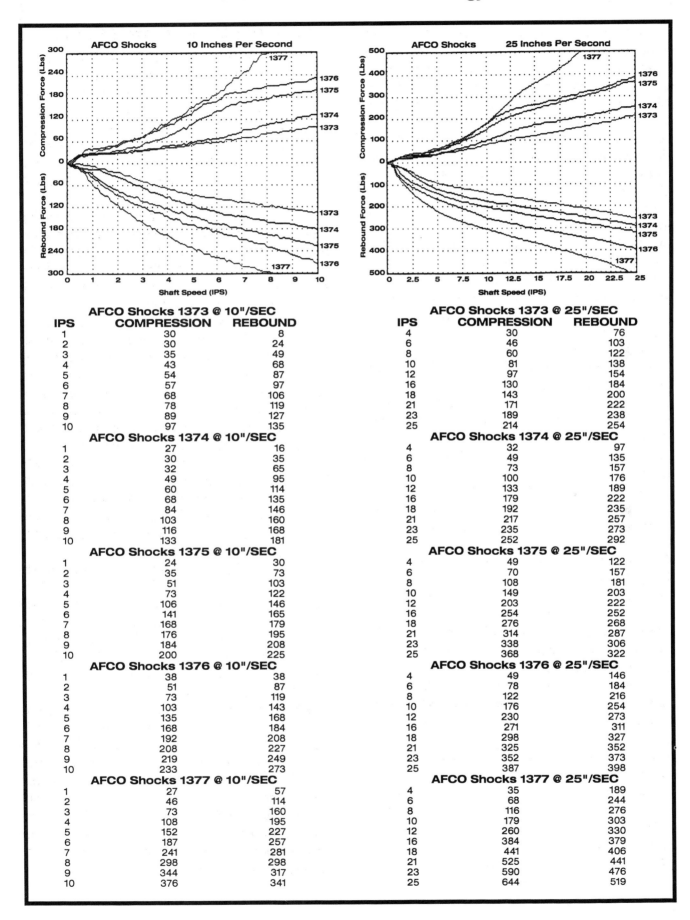

AFCO Shocks 1373 @ 10"/SEC

IPS	COMPRESSION	REBOUND
1	30	8
2	30	24
3	35	49
4	43	68
5	54	87
6	57	97
7	68	106
8	78	119
9	89	127
10	97	135

AFCO Shocks 1374 @ 10"/SEC

IPS	COMPRESSION	REBOUND
1	27	16
2	30	35
3	32	65
4	49	95
5	60	114
6	68	135
7	84	146
8	103	160
9	116	168
10	133	181

AFCO Shocks 1375 @ 10"/SEC

IPS	COMPRESSION	REBOUND
1	24	30
2	35	73
3	51	103
4	73	122
5	106	146
6	141	165
7	168	179
8	176	195
9	184	208
10	200	225

AFCO Shocks 1376 @ 10"/SEC

IPS	COMPRESSION	REBOUND
1	38	38
2	51	87
3	73	119
4	103	143
5	135	168
6	168	184
7	192	208
8	208	227
9	219	249
10	233	273

AFCO Shocks 1377 @ 10"/SEC

IPS	COMPRESSION	REBOUND
1	27	57
2	46	114
3	73	160
4	108	195
5	152	227
6	187	257
7	241	281
8	298	298
9	344	317
10	376	341

AFCO Shocks 1373 @ 25"/SEC

IPS	COMPRESSION	REBOUND
4	30	76
6	46	103
8	60	122
10	81	138
12	97	154
16	130	184
18	143	200
21	171	222
23	189	238
25	214	254

AFCO Shocks 1374 @ 25"/SEC

IPS	COMPRESSION	REBOUND
4	32	97
6	49	135
8	73	157
10	100	176
12	133	189
16	179	222
18	192	235
21	217	257
23	235	273
25	252	292

AFCO Shocks 1375 @ 25"/SEC

IPS	COMPRESSION	REBOUND
4	49	122
6	70	157
8	108	181
10	149	203
12	203	222
16	254	252
18	276	268
21	314	287
23	338	306
25	368	322

AFCO Shocks 1376 @ 25"/SEC

IPS	COMPRESSION	REBOUND
4	49	146
6	78	184
8	122	216
10	176	254
12	230	273
16	271	311
18	298	327
21	325	352
23	352	373
25	387	398

AFCO Shocks 1377 @ 25"/SEC

IPS	COMPRESSION	REBOUND
4	35	189
6	68	244
8	116	276
10	179	303
12	260	330
16	384	379
18	441	406
21	525	441
23	590	476
25	644	519

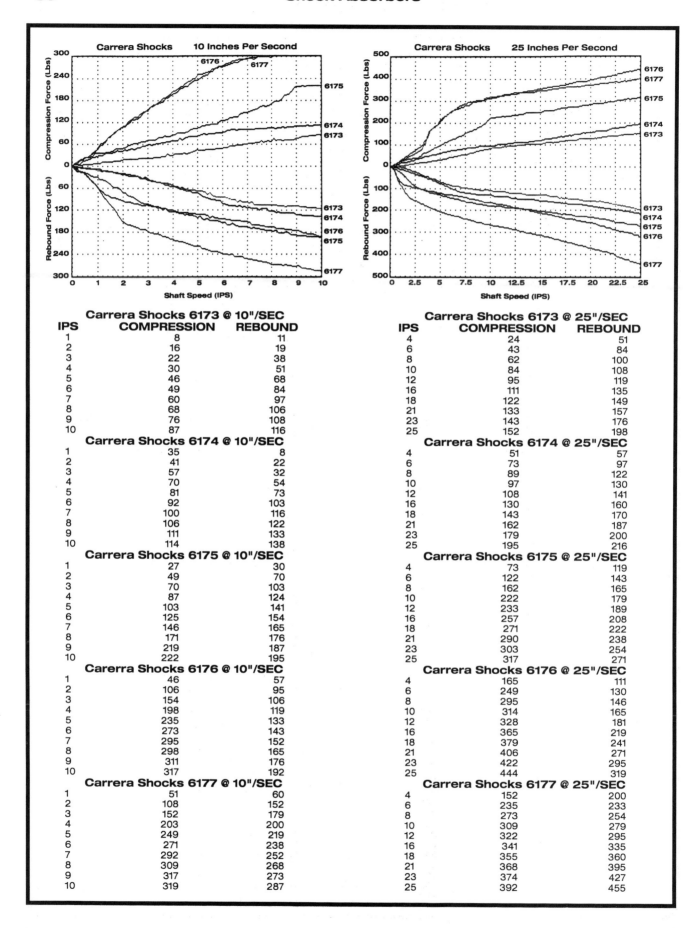

Carrera Shocks 6173 @ 10"/SEC

IPS	COMPRESSION	REBOUND
1	8	11
2	16	19
3	22	38
4	30	51
5	46	68
6	49	84
7	60	97
8	68	106
9	76	108
10	87	116

Carrera Shocks 6174 @ 10"/SEC

IPS	COMPRESSION	REBOUND
1	35	8
2	41	22
3	57	32
4	70	54
5	81	73
6	92	103
7	100	116
8	106	122
9	111	133
10	114	138

Carrera Shocks 6175 @ 10"/SEC

IPS	COMPRESSION	REBOUND
1	27	30
2	49	70
3	70	103
4	87	124
5	103	141
6	125	154
7	146	165
8	171	176
9	219	187
10	222	195

Carrera Shocks 6176 @ 10"/SEC

IPS	COMPRESSION	REBOUND
1	46	57
2	106	95
3	154	106
4	198	119
5	235	133
6	273	143
7	295	152
8	298	165
9	311	176
10	317	192

Carrera Shocks 6177 @ 10"/SEC

IPS	COMPRESSION	REBOUND
1	51	60
2	108	152
3	152	179
4	203	200
5	249	219
6	271	238
7	292	252
8	309	268
9	317	273
10	319	287

Carrera Shocks 6173 @ 25"/SEC

IPS	COMPRESSION	REBOUND
4	24	51
6	43	84
8	62	100
10	84	108
12	95	119
16	111	135
18	122	149
21	133	157
23	143	176
25	152	198

Carrera Shocks 6174 @ 25"/SEC

IPS	COMPRESSION	REBOUND
4	51	57
6	73	97
8	89	122
10	97	130
12	108	141
16	130	160
18	143	170
21	162	187
23	179	200
25	195	216

Carrera Shocks 6175 @ 25"/SEC

IPS	COMPRESSION	REBOUND
4	73	119
6	122	143
8	162	165
10	222	179
12	233	189
16	257	208
18	271	222
21	290	238
23	303	254
25	317	271

Carrera Shocks 6176 @ 25"/SEC

IPS	COMPRESSION	REBOUND
4	165	111
6	249	130
8	295	146
10	314	165
12	328	181
16	365	219
18	379	241
21	406	271
23	422	295
25	444	319

Carrera Shocks 6177 @ 25"/SEC

IPS	COMPRESSION	REBOUND
4	152	200
6	235	233
8	273	254
10	309	279
12	322	295
16	341	335
18	355	360
21	368	395
23	374	427
25	392	455

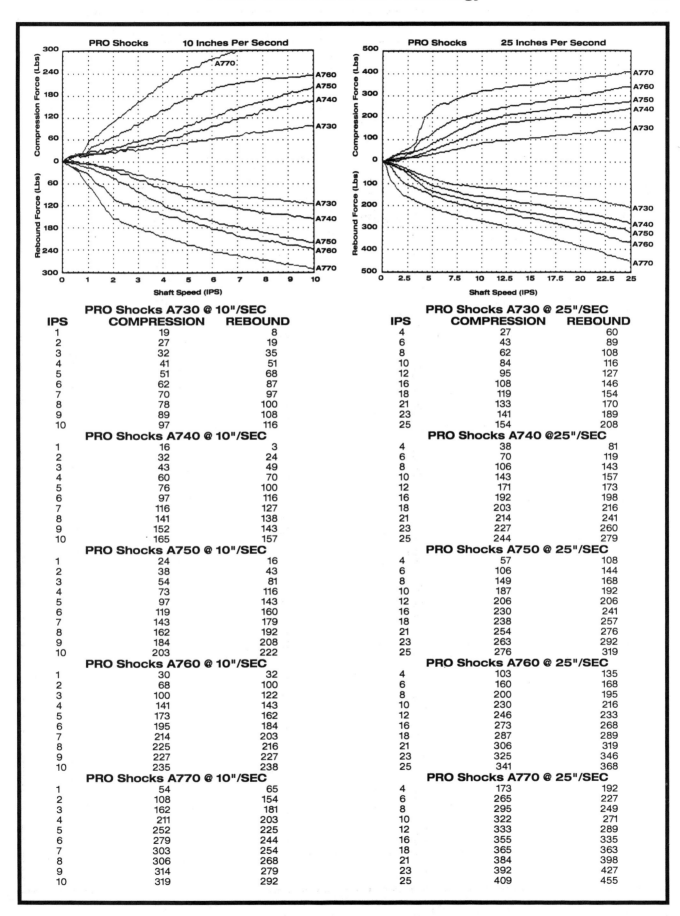

PRO Shocks A730 @ 10"/SEC

IPS	COMPRESSION	REBOUND
1	19	8
2	27	19
3	32	35
4	41	51
5	51	68
6	62	87
7	70	97
8	78	100
9	89	108
10	97	116

PRO Shocks A740 @ 10"/SEC

1	16	3
2	32	24
3	43	49
4	60	70
5	76	100
6	97	116
7	116	127
8	141	138
9	152	143
10	165	157

PRO Shocks A750 @ 10"/SEC

1	24	16
2	38	43
3	54	81
4	73	116
5	97	143
6	119	160
7	143	179
8	162	192
9	184	208
10	203	222

PRO Shocks A760 @ 10"/SEC

1	30	32
2	68	100
3	100	122
4	141	143
5	173	162
6	195	184
7	214	203
8	225	216
9	227	227
10	235	238

PRO Shocks A770 @ 10"/SEC

1	54	65
2	108	154
3	162	181
4	211	203
5	252	225
6	279	244
7	303	254
8	306	268
9	314	279
10	319	292

PRO Shocks A730 @ 25"/SEC

IPS	COMPRESSION	REBOUND
4	27	60
6	43	89
8	62	108
10	84	116
12	95	127
16	108	146
18	119	154
21	133	170
23	141	189
25	154	208

PRO Shocks A740 @25"/SEC

4	38	81
6	70	119
8	106	143
10	143	157
12	171	173
16	192	198
18	203	216
21	214	241
23	227	260
25	244	279

PRO Shocks A750 @ 25"/SEC

4	57	108
6	106	144
8	149	168
10	187	192
12	206	206
16	230	241
18	238	257
21	254	276
23	263	292
25	276	319

PRO Shocks A760 @ 25"/SEC

4	103	135
6	160	168
8	200	195
10	230	216
12	246	233
16	273	268
18	287	289
21	306	319
23	325	346
25	341	368

PRO Shocks A770 @ 25"/SEC

4	173	192
6	265	227
8	295	249
10	322	271
12	333	289
16	355	335
18	365	363
21	384	398
23	392	427
25	409	455

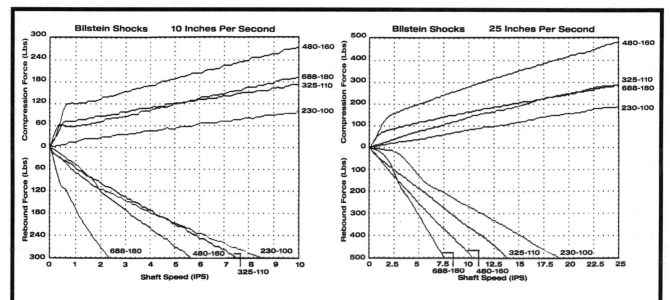

Bilstein Shocks 230 @ 10"/SEC

IPS	COMPRESSION	REBOUND
1	14	46
2	27	116
3	35	146
4	46	179
5	51	206
6	65	235
7	73	260
8	81	289
9	89	319
10	100	344

Bilstein Shocks 325 @ 10"/SEC

IPS	COMPRESSION	REBOUND
1	73	57
2	87	97
3	97	135
4	108	170
5	119	208
6	133	246
7	141	279
8	152	317
9	162	354
10	173	398

Bilstein Shocks 480 @ 10"/SEC

IPS	COMPRESSION	REBOUND
1	119	70
2	133	122
3	154	170
4	171	219
5	187	271
6	203	317
7	225	365
8	238	414
9	257	455
10	273	506

Bilstein Shocks 688 @ 10"/SEC

IPS	COMPRESSION	REBOUND
1	57	170
2	70	273
3	87	354
4	100	427
5	122	498
6	135	571
7	146	644
8	165	725
9	176	793
10	192	874

Bilstein Shocks 230 @ 25"/SEC

IPS	COMPRESSION	REBOUND
4	32	73
6	46	170
8	62	219
10	81	271
12	95	319
14	108	368
16	127	425
18	143	476
21	160	552
23	181	617
25	187	676

Bilstein Shocks 325 @ 25"/SEC

IPS	COMPRESSION	REBOUND
4	106	152
6	127	222
8	146	292
10	162	363
12	181	425
16	211	593
18	230	674
21	257	787
23	273	860
25	287	923

Bilstein Shocks 480 @ 25"/SEC

IPS	COMPRESSION	REBOUND
4	184	203
6	217	300
8	249	390
10	279	482
12	311	595
16	365	822
18	392	925
21	430	1069
23	457	1169
25	485	1255

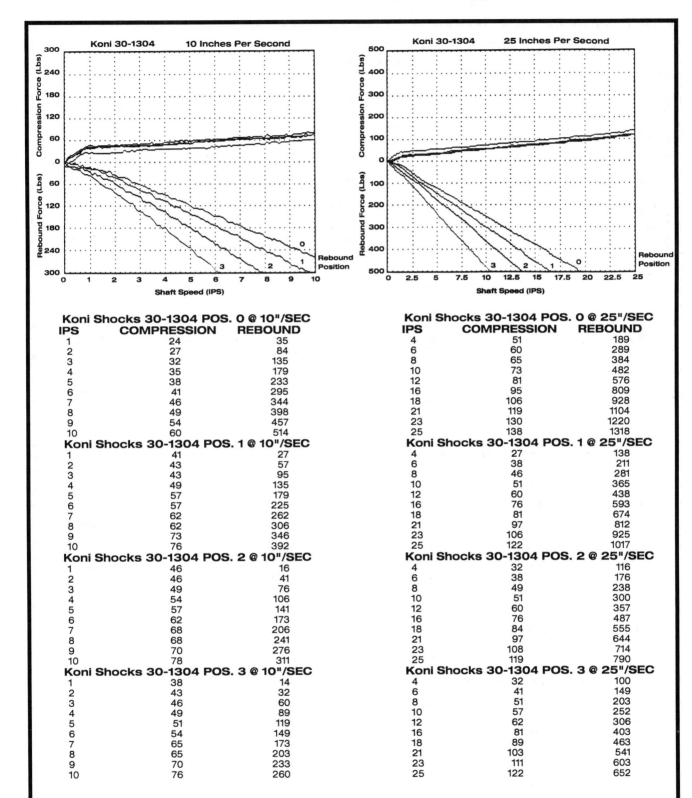

Koni Shocks 30-1304 POS. 0 @ 10"/SEC

IPS	COMPRESSION	REBOUND
1	24	35
2	27	84
3	32	135
4	35	179
5	38	233
6	41	295
7	46	344
8	49	398
9	54	457
10	60	514

Koni Shocks 30-1304 POS. 1 @ 10"/SEC

1	41	27
2	43	57
3	43	95
4	49	135
5	57	179
6	57	225
7	62	262
8	62	306
9	73	346
10	76	392

Koni Shocks 30-1304 POS. 2 @ 10"/SEC

1	46	16
2	46	41
3	49	76
4	54	106
5	57	141
6	62	173
7	68	206
8	68	241
9	70	276
10	78	311

Koni Shocks 30-1304 POS. 3 @ 10"/SEC

1	38	14
2	43	32
3	46	60
4	49	89
5	51	119
6	54	149
7	65	173
8	65	203
9	70	233
10	76	260

Koni Shocks 30-1304 POS. 0 @ 25"/SEC

IPS	COMPRESSION	REBOUND
4	51	189
6	60	289
8	65	384
10	73	482
12	81	576
16	95	809
18	106	928
21	119	1104
23	130	1220
25	138	1318

Koni Shocks 30-1304 POS. 1 @ 25"/SEC

4	27	138
6	38	211
8	46	281
10	51	365
12	60	438
16	76	593
18	81	674
21	97	812
23	106	925
25	122	1017

Koni Shocks 30-1304 POS. 2 @ 25"/SEC

4	32	116
6	38	176
8	49	238
10	51	300
12	60	357
16	76	487
18	84	555
21	97	644
23	108	714
25	119	790

Koni Shocks 30-1304 POS. 3 @ 25"/SEC

4	32	100
6	41	149
8	51	203
10	57	252
12	62	306
16	81	403
18	89	463
21	103	541
23	111	603
25	122	652

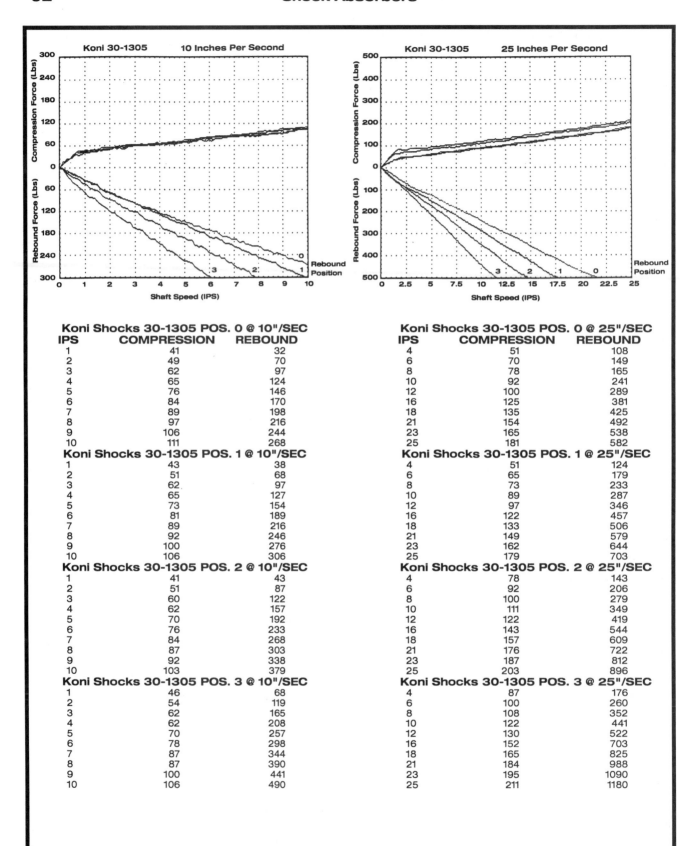

Koni Shocks 30-1305 POS. 0 @ 10"/SEC

IPS	COMPRESSION	REBOUND
1	41	32
2	49	70
3	62	97
4	65	124
5	76	146
6	84	170
7	89	198
8	97	216
9	106	244
10	111	268

Koni Shocks 30-1305 POS. 1 @ 10"/SEC

IPS	COMPRESSION	REBOUND
1	43	38
2	51	68
3	62	97
4	65	127
5	73	154
6	81	189
7	89	216
8	92	246
9	100	276
10	106	306

Koni Shocks 30-1305 POS. 2 @ 10"/SEC

IPS	COMPRESSION	REBOUND
1	41	43
2	51	87
3	60	122
4	62	157
5	70	192
6	76	233
7	84	268
8	87	303
9	92	338
10	103	379

Koni Shocks 30-1305 POS. 3 @ 10"/SEC

IPS	COMPRESSION	REBOUND
1	46	68
2	54	119
3	62	165
4	62	208
5	70	257
6	78	298
7	87	344
8	87	390
9	100	441
10	106	490

Koni Shocks 30-1305 POS. 0 @ 25"/SEC

IPS	COMPRESSION	REBOUND
4	51	108
6	70	149
8	78	165
10	92	241
12	100	289
16	125	381
18	135	425
21	154	492
23	165	538
25	181	582

Koni Shocks 30-1305 POS. 1 @ 25"/SEC

IPS	COMPRESSION	REBOUND
4	51	124
6	65	179
8	73	233
10	89	287
12	97	346
16	122	457
18	133	506
21	149	579
23	162	644
25	179	703

Koni Shocks 30-1305 POS. 2 @ 25"/SEC

IPS	COMPRESSION	REBOUND
4	78	143
6	92	206
8	100	279
10	111	349
12	122	419
16	143	544
18	157	609
21	176	722
23	187	812
25	203	896

Koni Shocks 30-1305 POS. 3 @ 25"/SEC

IPS	COMPRESSION	REBOUND
4	87	176
6	100	260
8	108	352
10	122	441
12	130	522
16	152	703
18	165	825
21	184	988
23	195	1090
25	211	1180

How To Choose The Shocks For Your Application

Looking at the shock dyno graphs of different shock absorbers can give you a lot of comparative information. There are two areas a racer should look at. One is the 0 to 10 inches per second range. This is the critical range for control of body roll and pitch. From 3 to 5"/sec. is the low speed control for weight transfer during corner entry and the middle of the turns. The 8 to 10"/sec. range relates to faster weight transfer control such as heavy braking, corner exit and small bump and rut control.

A graph of the 0 to 10"/sec. control range should rise quickly and smoothly, then the acceleration of this curve should taper off. It should then transition into the second critical area — the 12 to 14 inches per second range. Here it should flatten out. The 12 to 14 inches per second velocity is the control range for small bumps and ruts. The shocks cannot be stiff in compression here, or they won't allow the tires to follow the contour of the track over minor bumps, ruts and undulations. Too high a damping force in the mid ranges will cause the car to "wash out" over rough areas of the track.

Variations In Shock Absorbers

The accompanying shock dyno charts are representative of the part numbers for the shocks specified. But you should be aware that not all shocks are made the same. There are manufacturing tolerances involved when building and assembling the precise parts that go into shocks. Chances are that if you pull four of the same part number of any brand of shock off the shelf, and dyno them, you will find small variations in the compression and rebound damping rates between each. This is normal. You will also find that double tube shock absorbers which are called 50/50 (same rating in compression and rebound) are not truly 50/50. They will have slightly more rebound control at middle and high speeds. The rebound is usually a little stiffer so it can control the compressed energy of the spring when it rebounds.

Many top racers will ask their shock manufacturer to provide (for a small extra fee) a shock dyno chart along with their shocks. This way they will know for sure what the compression and rebound rates are for each shock. This helps make sure that when they are searching for a shock that has a lot of rebound control, for instance, they know exactly how much each one has. It also provides a base of comparison when the shock has been used for a season. Almost every major shock manufacturer provides a dyno testing service for racers on both new and used shocks. They will also test shocks from other manufacturers for a slightly higher fee.

Every racer should be aware that the valving and mechanicals inside of shocks change over the years. For instance, a 95 shock that you buy today is definitely not the same shock that you purchased 7 years ago, even it carries the same part number. The product that you can buy today is a much more sophisticated piece.

The reason is that all shock manufacturers continue to do research and develop their products, and the shocks get better over time. New innovations and advancements are incorporated into the shocks as they are manufactured. But the part numbers are kept the same because they still reflect the true application of the shock as far as stroke length and damping force valve coding.

The bottom line is that if you are still racing with shocks that are 4 or 5 years old, buy some new shocks. You will be surprised at the difference in control the new shocks offer.

Chapter

5

The Braking System

Why concentrate on a competitive braking system for a race car? Very simply, stopping a race car quicker in a shorter distance can be one of the biggest advantages a racer can have. Frankly, very few racers really concentrate on brakes as an area to improve speed. They would rather concentrate on bite off a corner, or horsepower for the straightaway. But a driver with a good braking system can overcome those other advantages on turn entry.

Selecting the proper brake system components for your particular race car and application will help the car go faster. This is because when the braking system and components are optimized, the driver can go deeper into the turns while the system will allow him to do that lap after lap without experiencing brake fade.

Braking System Hydraulic Basics

How the master cylinder feeds pressure through the braking system is a basic lesson in hydraulics.

The system begins with an actuating force fed into the master cylinders. This force is the driver's foot on the brake pedal. The most force a driver can generate is between 150 and 200 pounds. But most racing brake applications are about 75 to 100 pounds of force.

The driver is also limited to about six inches in the amount of pedal travel he can apply (depending on the driver's leg length, closeness to the pedal and leg angle). The ability of the driver to apply force to the pedal is determined by body restraints (seat belt and shoulder harness), and the angle of the hip, knee and ankle joints. The more these three joints are in a straight line, the easier it is to create higher forces on the pedal.

The force from the foot is multiplied by the leverage of the brake pedal lever on the master cylinder pushrod. The most common pedal leverage ratio used on pavement stock cars is 6 to 1. This leverage ratio multiplies the driver's input force into the master cylinder. If 100 pounds of force is applied to the brake pedal and the leverage ratio is 6 to 1, the input force into the master cylinder is 600 pounds (100 times 6). If the pedal ratio is increased, the output force will increase, but the pedal travel will also increase.

The master cylinder bore size (and thus its square inch piston area) is the next consideration. A common master cylinder size is one inch. This gives an area of 0.785 square inches (area is found by the formula $A = .785$ times D squared, where D is the piston diameter).

The brake system output pressure (measured in pounds per square inch or PSI) generated by the master cylinder is calculated by the formula: F divided by A, where F is force (the leg input in pounds)

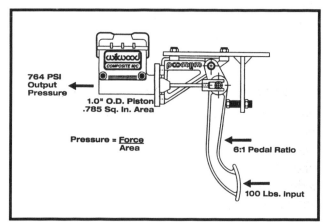

Image courtesy of Wilwood Engineering

When the master cylinders are mounted higher, they are mounted away from the heat of the header exit, and dirt and tire rubber are not constantly collecting on the master cylinders.

and A is area of the piston (calculated in above paragraph). For example, if a 1-inch diameter master cylinder is used, and the driver inputs 100 pounds with a 6 to 1 ratio pedal, the force is 600 pounds (6 times 100) and the area is .785, so 600 divided by .785 is 764 PSI system pressure. If a 7/8-inch master cylinder piston was used instead, the system pressure would be 998 PSI.

From this knowledge, it is easy to see that the smaller the master cylinder bore, the more pressure that is supplied to the calipers at the wheels. So why not use a smaller master cylinder piston diameter, such as 3/4 or 1/2-inch? There are two reasons against it: 1) It requires more pedal travel for a given fluid displacement, and 2) Higher line pressure aggravates line and caliper expansion and thus uses up more fluid. If a braking system is sufficiently stiff (meaning no or little line expansion or caliper deflection) that the driver is only using a small part of the total pedal travel available, then a master cylinder with a smaller bore would reduce the pedal effort (force) required by the driver to stop the car.

Brake Pedal Arrangement

Pedals can be mounted in a hanging position or as a floor mount. The hanging position offers several advantages. When the master cylinders are mounted higher, they are mounted away from the heat of the header exit, and dirt and tire rubber are not constantly collecting on the master cylinders. The higher

mounting also simplifies brake bleeding. When the master cylinders are at the highest point in the braking system, all fluid drains downhill. This aids in being able to bleed all the air out of the system. When the master cylinders are mounted at the same level as, or lower than, the calipers, brake fluid will drain back to the master cylinder reservoir, leaving a vacuum in the caliper and causing a slight caliper piston retraction. This requires the driver to pump the pedal several times to move fluid back into the caliper. To prevent this, use an inline 2 PSI residual pressure valve in front of the master cylinders.

Pedal Ratio

Pedal ratio simply compares the distance the pedal travels to the distance that the actuating rod to the balance bar (or master cylinder) travels. The formula

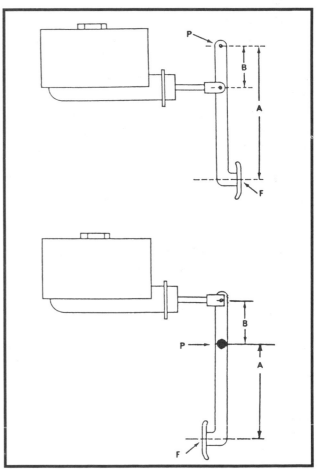

Pedal ratio is "A" divided by "B". "P" is the pivot point of the lever arm, and "F" is the input force. Pedal ratio is a method of gaining more brake pressure with mechanical leverage by multiplying the driver's input force. As the pedal ratio increases, pedal pressure can be reduced, but the pedal stroke increases.

Steel brake lines should run consistently downhill from the master cylinders to prevent trapped air in the lines. Bends in the lines should be made with a brake tubing bender. Note the use of rubber-lined hold-down clamps which help insulate the lines from vibration.

for determining the ratio is A divided by B (see diagram). If A = 6 inches and B = 1 inch, the ratio is 6 to 1. On racing brake assemblies, the pedal ratio is generally 6 to 1 or 7 to 1. A higher ratio reduces pedal effort but increases pedal travel. A lower ratio increases effort but reduces travel. A driver is most comfortable with a brake pedal feel that is very firm and has minimum pedal travel. That gives him the confidence to know that his brakes are there and helps him to modulate the braking application.

The Brake Lines

The brake lines function as the blood system of the braking system, and just as in the human body, there are several pertinent factors required to permit good circulation.

For all fixed lines in the system, use automotive double wall (Bundyweld) tubing with 3/16-inch outside diameter. Any larger diameter tubing may expand and create a spongy pedal because of the high operating pressure of a racing disc brake system. Never use copper tubing — it is too soft.

You will notice that throughout this chapter the rigidity of the braking system is mentioned over and over. The brake lines play a big part here. This is important if you want the force you impart to the master cylinder to be delivered to the caliper pistons. If you use that force in expanding the brake lines or

calipers instead of applying force to the brake pads, you cannot expect to stop the vehicle effectively, and the driver will merely use up all the available pedal travel. The pedal will bottom-out on the floor or the master cylinder piston will bottom in its bore. By making the system as rigid as possible, you are going to get the work to the pads, where it belongs.

The layout of the solid tubing running from the master cylinder to each caliper should be carefully designed to prevent the presence of unnecessary bends and loops which can trap air. And, the lines should run consistently downhill from the master cylinder to the breaking junction of the flex lines to aid in line bleeding.

Use a brake line tubing bender (available at parts stores or tool stores) to make good, neat tubing bends. Bending by hand or over a mold will produce crimps in the tubing.

Many people believe it is possible to proportion the braking force between the front and the rear brakes by installing larger diameter lines to the front wheels than the rear. This is NOT true. Pressure is independent of volume or area, so a 3/16-inch line to the front wheels will deliver the same pressure (PSI) as a 1/8-inch line to the rear wheels.

Never route brake lines near any source of heat, such as headers or exhaust pipes. Also, never route the brake lines in such a way as to have a loop or kink or high point where air can get trapped and defy bleeding techniques.

Steel brake lines should be secured to the chassis in several places along their run with rubber-lined Adel clamps. These aircraft-designed clamps prevent the transfer of vibration to the brake lines which could otherwise fracture the line material. All tee fittings along the steel tubing run should be secured to the chassis with a bracket.

At the connecting points of steel brake tubing, the fitting flares should be standard automotive double flares to prevent tubing fractures in the flares. And always cut the steel tubing with a tubing cutter, not a hack saw.

Swagelok Brake Line Fittings

Are you tired of spending your time flaring tubing to fit in brake fittings? Try using the Swagelok tube fittings. All you have to do is cut the brake line to size and insert the end into the Swagelok fitting, and then tighten it down according to instructions. It's

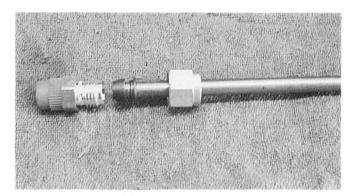

The Swagelok fitting uses an internal compression fitting to eliminate the need for flaring brake tubing.

simple and quick. The fitting has an internal compression fitting with a collar behind it to add extra force on the line as the nut is tightened. The fittings are available in a variety of configurations.

Flex Lines

At the point where the flex lines connect to the steel brake lines, the steel line must be secured to the chassis by a solid bracket mount. Otherwise, the input movements from the flex lines will also move the steel line, eventually causing a failure.

Only stainless steel braided teflon-lined flex hoses should be used to carry fluid from the steel lines to the calipers with the assurance that little pedal travel or work effort will be lost through line expansion. There are many brands of high quality plumbing lines available, but one of the best is made by Earl's Performance Products. Earl's manufactures Speed-Flex steel braided hose specifically intended for use as brake flex lines. It uses an extruded Teflon inner liner, with a tightly woven high tensile strength stainless steel outer braid to protect against brake line swelling during braking applications which improves brake pedal feel and firmness for the driver. Use the "-3" (dash three) size hose, which has a 0.125-inch inside diameter, and 2,000 PSI maximum operating pressure.

When you buy flex line, be sure you know what you are getting — there are a lot of imitations out there that look like the real thing. There is also neoprene-lined steel braided hose available, but do not use it. Neoprene is not designed for use in high pressure applications such as the braking system.

To determine the amount of flex line to use at each wheel, jack the car up and let the wheel rest at full droop. Then install a line of adequate length which

The breaking junction between the steel brake lines and the flex lines should be secured to the chassis with a bracket.

Always loop the front brake flex line excessive length. This acts as a spring to pull the line away from the wheel inner side.

is not tight. For front wheels, steer the wheel hard in both directions as well as setting at full droop to determine the maximum length of line required. Always loop the front brake flex line excessive length. This acts as a spring to pull the line away from the wheel inner side.

Never assume that any brake lines or hoses you purchase are clean. Pour clean brake fluid through them just before final installation to be sure the smallest amounts of moisture, dirt and metal cuttings are removed. And, make sure the fittings into the calipers are not cross threaded.

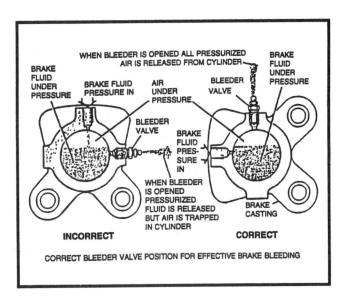

WHEN BLEEDER IS OPENED ALL PRESSURIZED AIR IS RELEASED FROM CYLINDER

BRAKE FLUID UNDER PRESSURE

BRAKE FLUID PRESSURE IN

AIR UNDER PRESSURE

BLEEDER VALVE

BRAKE FLUID UNDER PRESSURE

BLEEDER VALVE

BRAKE FLUID PRESSURE IN

WHEN BLEEDER IS OPENED PRESSURIZED FLUID IS RELEASED BUT AIR IS TRAPPED IN CYLINDER

BRAKE CASTING

INCORRECT   **CORRECT**

CORRECT BLEEDER VALVE POSITION FOR EFFECTIVE BRAKE BLEEDING

Brake Bleeding

Once the braking system has been installed, all that is required is to add fluid into the master cylinders and depress the brake pedal to distribute it into the system. As the fluid travels through the system, it will displace the air which filled it. When the fluid reaches the calipers, there will be a large amount of accumulated air which must be released. This is the reason for the bleeder screw, which should be fitted at the highest point of the brake caliper.

To bleed the system, open the bleeder screw with a box wrench, attach a hose over the fitting, and run the hose into a jar containing NEW brake fluid. Be sure the tube is well covered in the jar with fluid so air will not be sucked back into the system.

Start with the farthest wheel from the master cylinder, which is usually the right rear, and progressively work to the closest caliper. Have an assistant press the brake pedal to the floor. Air and fluid will escape in spurts together. Make sure your assistant returns the brake pedal back very slowly. After the brake pedal has been depressed, close the bleed screw before returning the pedal. This prevents air from being sucked back up into the system, and also tests the working quality of the master cylinder.

Never pump the brake pedal. This will aerate the fluid in the lines and you will never get the air out. Continue this process until all air has escaped and a solid column of fluid is observed. Then move to the next wheel. Be sure to check the fluid level in the master cylinder after each wheel has been bled.

Keep the master cylinder cap on while bleeding the brakes to avoid moisture contamination of the fluid.

With disc brake systems, air bubbles are difficult to dislodge sometimes after the initial installation of the calipers. Use a plastic or wooden mallet to tap the calipers gently to assist the air bubbles. Make sure the brake bleeder screws are positioned in an upward position when bleeding the brakes. This prevents the possibility of trapping air bubbles.

If calipers are unbolted from the suspension for bleeding, make sure that a piece of wood the appropriate thickness is inserted in place of the rotor so that the pistons are not over extended.

It is a good idea to bleed the brakes after a few practice runs have been made at the track, especially with a new or just-repaired system. It is also a good idea to bleed the brakes and flush a couple of cans of fresh fluid through the system after the heat races and before the main event.

Be sure to change the brake fluid in your race car on a weekly basis (or just before a racing event if you do not race weekly). Brake fluid absorbs moisture from the atmosphere, which lowers its boiling point. Humidity in the air, washing the car, and condensation in the system caused by heat cycling all contribute to water build-up in the fluid. Even piston seals, when they are extended and retracted, will allow a small amount of humidity into the fluid. When changing the fluid, flush the system with denatured alcohol, which absorbs water. Put a dye in the alcohol so when you add new brake fluid, you know that all of the alcohol has been flushed through the system.

Brake Fluids

Brake fluid transmits force through pressure from the driver's foot to the brake pads. The basic premise which allows this to happen is that the fluid is not compressible. Thus, it is able to transmit force.

The enemy of brake fluid, just like friction materials, is heat. If the fluid boils at any point in the system, or if the system leaks at any point, the fluid incompressibility is lost.

The important thing to keep in mind when purchasing brake fluids is that they are not all the same. The main ingredient is ethylene glycol, which has a lubricating capability for the system's rubber parts and is not highly susceptible to boiling. The two critical elements to consider when choosing a brake fluid are the dry boiling point of the fluid and the viscosity of the fluid.

Brake fluids in the United States are rated by the Department of Transportation (DOT). The DOT standards set minimum specifications for wet and dry boiling points of brake fluid. ALWAYS use a brake fluid that at least conforms to the DOT 3 standard, and much preferably exceeds it. The dry boiling point is the brake fluid in its pure state with no moisture contamination. The wet boiling point is after it has been saturated with moisture. The wet boiling point is the critical number to look at. The dry boiling point for high performance racing DOT 3 fluid is 572° F, while its wet boiling point is 284° F. DOT 4 fluid is a higher standard, having a minimum wet boiling point of 311° F.

Brake fluids which meet the DOT 3 requirements are Wilwood's Hi-Temp 570, Sierra Racing Products Z-10, and Motul 570.

Many racers are now using DOT 4 fluid because of its higher wet boiling point. Two fluids which meet these standards are AP 600 and Castrol SRF. The AP 600 has a wet boiling point of 311° F, while the Castrol SRF has a 518° F wet boiling point.

Glycol-based brake fluids are highly moisture absorbent when exposed to the atmosphere. Because of this, it is important to buy brake fluid in small containers, such as a 12-ounce can, so the contents can be used up all at one time. The fluid in both the can and the master cylinder should be exposed to the atmosphere for the shortest possible time to prevent moisture absorption. The container in which the fluid is stored should be kept tightly sealed until the fluid is used. If you have any fluid left in a can after filling the system, throw it away. Don't save it and take a chance of using contaminated fluid. It is also a good idea to keep the master cylinder full, and be sure its cover has an expandable rubber boot which will take up air space as the fluid level decreases.

You should be aware that there is a DOT 5 standard for brake fluids. This is strictly for silicone brake fluids. These fluids have a higher boiling point, and as such have been promoted as a great racing brake fluid. Be aware, though, that it is not recommended for use in a race car. It contains several inherent problems: 1) It aerates easily, and 2) it is highly expansive under higher temperatures, causing more compressibility. Don't use it.

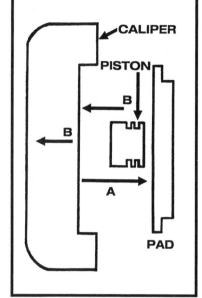

How caliper deflection can happen: Force from the hydraulic system is in direction A. The reaction to the force is in direction B, which tries to deflect the caliper housing.

Calipers

The major item to consider in selecting components is caliper housing deflection. To explain why deflection is so detrimental to the caliper, refer to the accompanying drawing. Remember that basic law of physics, "For every action, there is an opposite and equal reaction?" That certainly applies in the caliper housing. In the drawing, "A" refers to the initial action of the piston being forced against the pad, which in turn is forced against the disc. "B" represents the opposite and equal reaction. If the caliper housing was flexible enough, the system could use the rotor as the anchoring or back up point, and the fluid pressure would then apply force against the housing. This would deflect the caliper using all available pedal travel and deliver little if any stopping ability.

Every disc brake manufacturer is concerned about caliper flex, which is the real enemy of producing proper braking force. And they all approach the battle a little differently – different caliper materials, different bridge bolting patterns, external stiffening ribs on the housing, and cast versus billet housing material.

Most aftermarket caliper designs use a stiffening rib or ribs across the caliper body to increase body strength and rigidity, which combats flex.

Aftermarket calipers are made from cast aluminum or billet aluminum. One type of material is not necessarily better than the other. It depends upon the

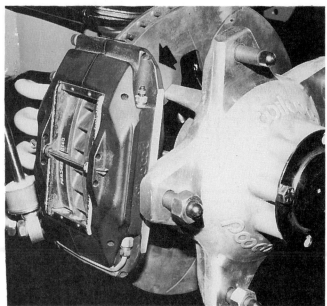

Most aftermarket caliper designs use a stiffening rib or ribs across the caliper body to increase body strength and rigidity, which combats flex.

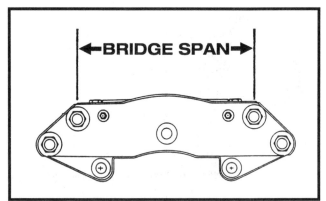

The bridge distance is the span between clamping bolts.

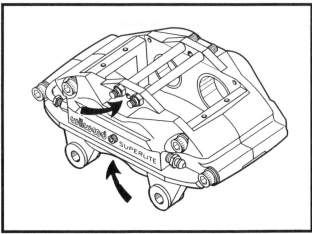

Thicker material in the housing between the bridge bolts and external stiffening ribs, as well as double center bridge bolts, add tremendously to caliper body strength and rigidity. Image courtesy of Wilwood Engineering.

application and the caliper design. A cast caliper can be shaped in ways that would be difficult to machine from billet aluminum. Good casting techniques and materials can produce a very rigid casting which gives a very high modulus of elasticity for ultimate stiffness. A billet material has a higher tensile strength than a cast material, but because the billet stock's grain all runs in one direction, it can flex and return to its original shape.

The bottom line is that both cast aluminum and billet aluminum are excellent materials for a caliper. What creates the ultimate in stiffness is the design of the part – how it is shaped and how stiffening design elements are engineered into it, such as bridge span and bridge bolting.

The bridge distance is the span between clamping bolts. The longer this span, the thicker the caliper housing material must be to resist deflection. Or, the longer this distance, the greater the deflection is going to be. So, the piston or pistons should be placed as close to the caliper clamping bolts as possible. Thicker material in the housing between the bridge bolts and external stiffening ribs help combat deflection. And, double center bridge bolts add tremendously to body strength and rigidity.

The question of bridge span is closely related to the piston size and arrangement. The best arrangement of pistons is two in a row on each side of the rotor. This spreads the force loading closer to the

clamping bolts, putting it in combined tension and bending against the bolts, reducing caliper deflection.

Heat conductivity from the rotor into the caliper and fluid is also a major concern. Fighting this heat problem starts with having the proper piston material. All top rated aftermarket calipers use stainless steel pistons. Stainless steel helps to resist thermal transfer.

Wilwood offers an optional 2-piece caliper piston. It is a stainless steel piston with a stainless steel liner inside, separated by a special insulator material. What this accomplished is a superior heat shielding because stainless steel is a good thermal barrier. This minimizes heat transfer through the pistons into the brake fluid and piston seals.

Most aftermarket manufacturers also incorporate some type of thermal barrier inside the caliper housing, usually in the form of a stainless steel heat shield

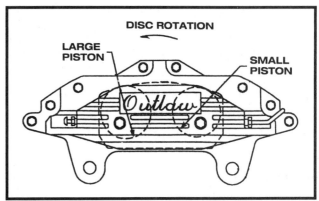

Differential bore calipers use two different sized pistons in the caliper to prevent brake pad taper. Image courtesy of Outlaw Disc Brakes.

These calipers incorporate an open bridge design which allows for better, less restricted, dissipation of hot air. Note how the caliper body ends are open around the bolts. Also note the stainless steel insulator inside the caliper.

or liner. This creates a thermal block that deflects heat from the rotor away from the caliper housing, helping it and the fluid to operate cooler.

A stainless steel insulator can also be placed between the rotor and its hat. This helps to minimize heat transfer from the rotor which can overheat wheels and wheel bearings.

A new type of caliper housing design called an open bridge also helps in keeping heat from transferring into the caliper housing. An open bridge design allows the hot air flow coming out of the rotor to escape the interior of the caliper without coming into contact with the caliper body. This greatly reduces conductive heating of the caliper body, and increases rotor air flow.

One other element that accelerates heat dissipation out of the caliper is a black anodized finish.

Differential Bore Calipers

Calipers with equally sized pistons normally create a tapered wear pattern on brake pads when they operate at a continuously high temperature range. That means the full surface area of the pad is not being applied against the rotor, so it is losing braking effectiveness.

What causes pad taper (from the leading edge to the trailing edge) is a byproduct of the extremely high operating temperatures found on the front brakes of short track stock cars. As operating temperatures

raise beyond 1,000° F, the pad material starts to release a gas. These gases build up and form a barrier between the pad and the rotor. The gases travel to the rear of the pad, and thus the front of the pad is subjected to more clamping force and it wears at a greater rate. To combat this, larger piston sizes are used at the rear edge of the pad for better clamping force there. Differential bore calipers use two different sized pistons in the caliper to prevent brake pad taper. A smaller diameter piston is used where the rotor enters the caliper so clamping force is decreased there. A larger diameter piston is placed at the rear of the caliper. This bore diameter configuration eliminates pad taper wear by differentiating the clamping force along the brake pad backing plate. Rotor grooves also allow the pads to vent their built-up gases.

Most racing brake manufacturers offer differential bore calipers in their line. All Outlaw disc brake calipers come standard with differential piston bores.

Differential bore calipers are not used on rear axle brakes because they do not encounter the extremely high operating temperatures that front brakes do.

Piston Seals

Piston seal performance is important to the braking system because the piston seals control retraction of the pistons. They must remain pliable so that the pistons function properly. When the seals are attacked by continuous extreme heat, the seals get hard. This can cause the pistons to stick, or under extreme heat conditions, they can fail and allow fluid to leak past them.

Caliper Maintenance

Disassemble the calipers on a monthly (or more regular) basis to inspect the caliper piston seals. Look for excessive seal wear and seal hardness. If the seals are questionable, replace them. This can prevent a catastrophic piston seal failure.

Also as a part of regular brake system maintenance, use brake cleaner to clean the interior of the caliper, the pistons and the piston bores, and replace all bleed screws and fittings.

Caliper Mounting

Brake calipers have to be mounted absolutely square with rotors to prevent excessive piston knock-back and uneven pad wear. A caliper that is not mounted square to a rotor will also introduce additional flex into the braking system, taking away from stopping ability. Many times caliper mounting brackets get bent. An easy way to check for this problem is to have somebody apply the brakes while you watch the caliper. The caliper should not try to square itself with the rotor – only the pistons and pads should move. If the caliper is not square with the rotor, use shims between the mounting bracket and the caliper ears to properly align it.

All caliper brackets should be stiff enough that they will not deflect under the heaviest of braking conditions. All caliper mounting bolts should be the highest quality hardware, and they should always be safety wired.

Floating Vs. Fixed Calipers

The fixed caliper is one which has pistons on each side of the rotor, squeezing the rotor from each side. Because hydraulic pressure is always self equalizing throughout the entire system, one piston cannot overcome any others. Nothing moves in the caliper except the pistons pushing the pads inward from each side.

The floating caliper was designed by passenger car manufacturers essentially to make the caliper less expensive to produce. It successfully applies the physics principle of "every action causes an equal and opposite reaction." Applying this principle, they eliminate pistons on one side of the rotor. The floating caliper is not solidly mounted but rather slides back and forth slightly on bushings. The piston travel on the primary side of the rotor is reacted by pulling on the housing so the pad on the secondary side is pulled tight against the rotor.

The floating caliper housing must be very rigid and of low deflection, or the entire principle behind it is lost, and so is braking ability. This is why the most effective full floating calipers are made of steel or cast iron — it offers a high modulus of elasticity.

If you use floating calipers, be sure to periodically check the condition of the bushings which the caliper floats on. If the sliding movement binds up on the bolt, there will not be a proper retraction, this causing piston drag.

Fixed calipers are definitely the type preferred for all-out racing conditions. Their advantages are greater pistons retraction control, and less flex.

Floating calipers are generally smaller and lighter. Their disadvantages: floating calipers tend to flex more, and their pad area is wider than and overlaps the piston face. This bends the pad and gives an uneven pad loading against the rotor. The multiple pistons found in fixed type calipers minimize this problem and give better pad clamping application against the rotor.

Two of the biggest problems with floating calipers are their tremendous frictional loss in contacting the outside pad, and their bridge design. On floating calipers, the bridge covers the top of the housing only — there is no material wrapping around the ends. This adds greatly to deflection and causes caliper clamping (after it heats up) even in the brake-released condition, robbing cool down time for the entire system.

Which Calipers And Master Cylinders To Use

Asphalt tracks demand the ultimate in race car stopping ability. Only the very best will do. There are several good companies making fine products — Outlaw, Sierra Racing Products, Wilwood, etc. Get all the catalogs, read them, talk to the manufacturers, and make your own choice. All make a wide selection of sizes. To help you choose the correct system for a typical 2,700 to 3,100 pound car, we have chosen to illustrate as an example the Wilwood Superlite III series calipers.

Brake System Specifications For Short Track Cars

Front
Calipers: Differential bore with 1.88" and 1.75" pistons (2 of each)

Rotors: 11.75" diameter, 1.25" width, 48 fin curved vane

Rear
Calipers: Four 1.38" pistons

Rotors: 11.75" diameter, 1.25" width, 32 fin curved vane

Master Cylinders
Front: .875" bore
Rear: .875" bore

Braking Force Calculation

On the front wheels, use the differential bore design which has two 1.88-inch O.D. pistons and

Braking Force Application

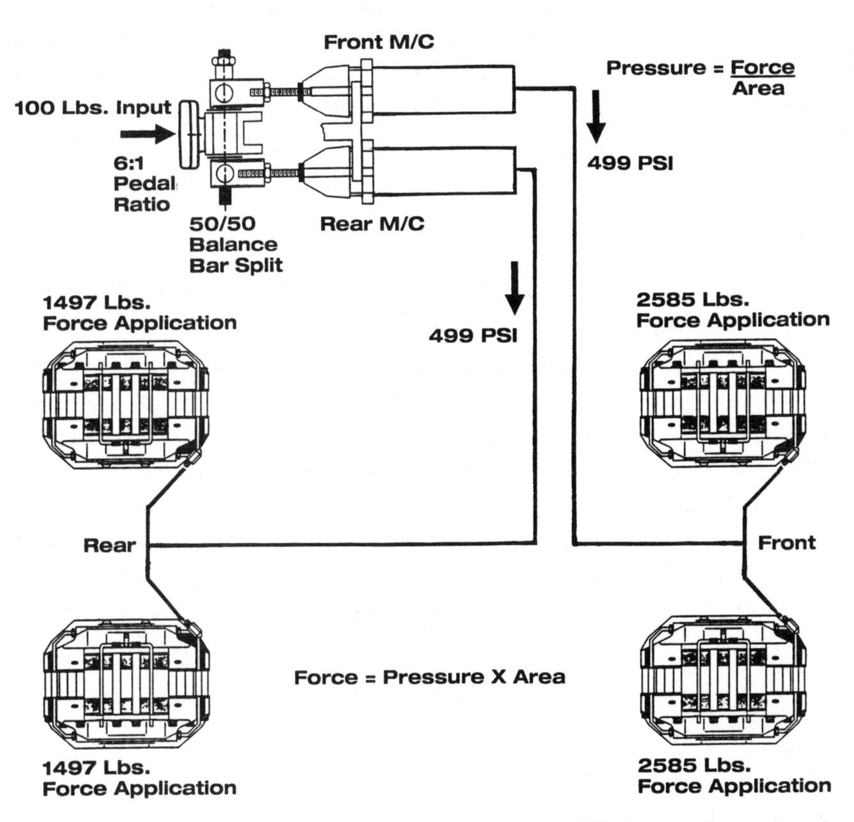

Front M/C

100 Lbs. Input

6:1 Pedal Ratio

50/50 Balance Bar Split

Rear M/C

Pressure = $\frac{Force}{Area}$

499 PSI

499 PSI

1497 Lbs. Force Application

2585 Lbs. Force Application

Rear

Front

Force = Pressure X Area

1497 Lbs. Force Application

2585 Lbs. Force Application

Images courtesy of Wilwood Engineering

two 1.75-inch O.D. pistons per caliper. On the rear wheels use Superlite IIA calipers with four 1.38-inch O.D. pistons per caliper. Use two 7/8-inch bore master cylinders, one for the front calipers and one for the rears. Use a 6 to 1 pedal ratio, and we will assume a 100-pound input into the pedal from the driver's foot.

First figure the force input into the master cylinder. 100 pounds times 6 (the pedal ratio) equals 600 pounds. Assume initially that the balance bar is set for equal force distribution between the front and rear master cylinders.

Next figure the master cylinder area, using $A = .785$ times D squared (A is area and D is piston diameter). In this case, A equals .785 times .875 squared, or .601 square inches for both master cylinders.

The system pressure is figured next, using Pressure = Force divided by Area. The total force is 600 pounds, and splitting that 50 percent to each master cylinder is 300 pounds for the front and 300 pounds at the rear, so 300 divided by .601 equals 499 PSI system pressure coming out of each master cylinder.

The caliper piston area is the next item to be computed, using Area = .785 times D squared times the number of caliper pistons.

For the front, .785 times 1.88 squared equals 2.78 square inches, and .785 times 1.75 squared equals 2.40 square inches, for a total of 5.18 square inches per caliper. Multiplied by two front calipers, this gives 10.36 square inches of piston area in the front calipers. (Note: A caliper's piston area is calculated by finding the total piston area from one side of the caliper only, not all four pistons.)

At the rear, .785 times 1.38 squared times 2 pistons equals 3.0 square inches. Multiply this by 2 calipers for a total of 6.0 square inches of piston area in the rear calipers

Finally the output force (which is the force exerted by the pistons to the pads) is found by multiplying the system pressure by the caliper piston area. For the front, 499 PSI times 10.36 square inches is 5,170 pounds of force. At the rear, 499 PSI times 6.0 square inches is 2,994 pounds of force.

To calculate what the braking force proportion is front to rear, add together the total front braking force and the total rear force. In this case, 5,170 plus 2,994 equals 8,164. Divide the front force by the total force. so 5,170 divided by 8,164 equals .63, or 63 percent of the total force distribution to the front brakes. This means that 37 percent goes to the rear brakes.

The braking system front-to-rear bias should be proportioned as closely as possible with caliper piston sizes. The proper component selection should provide at least 60 percent of the braking force coming from the front brakes, which in this example we have achieved. The fine tuning of the front-to-rear brake proportioning can come from adjustments with the balance bar. The balance bar adds just a fine tuning mechanism and is not meant to correct gross miscalculations in component choice.

These output force numbers shown above are very typical for a paved track stock car. If you are choosing components to design and engineer your own braking system, use the steps we have shown above to calculate the output force you will have at the caliper pistons. Your numbers should be close to the ones shown here.

Note: Many Limited Stock and Pro Stock racing association rules require that only OEM type of brake calipers can be used, or if aftermarket components are used, a 100-pound weight penalty is assessed to the car. If you face that situation, use the better aftermarket brakes and take the weight penalty. Your overall performance on the track will be much improved with superior brakes.

Some Tips For Working With GM-Style Calipers

Heat transfer from the pads into the brake fluid is always a constant problem with racing disc brakes. It is even more of a problem when using the smaller GM-based calipers. To *significantly* cut down the amount of heat transfer, use two pieces of .010-inch thick stainless steel sheet stock as a shim between the pistons and the back of each brake pad. Stainless steel has a low heat conductivity, and when two pieces of shim stock are used interfacing each other, the conductivity is cut way down.

Many racers in Limited Late Model or Modified classes are required to use stock calipers and rotors. For a paved track application, you need the biggest and best you can get on the front wheels. Use the Chevy Impala/1977-79 Cadillac Seville front calipers and rotors. The calipers use a 2.75-inch O.D. piston, and the rotors are 12 inches in diameter. For the rear, use the 1978 and later Chevelle/GM mid-size car calipers, which have 2.375-inch O.D. pistons.

Rotors

Understanding how a rotor works will help you to choose the correct rotors or your application. The stopping power of the caliper clamps the pads to the rotor, creating friction. The friction converts kinetic energy (moving energy) into heat energy. It is up to the rotor to dissipate the heat energy into the atmosphere and away from the brake fluid and seals. The type of rotor chosen to do this work depends on the demands placed on the braking system. A larger diameter, thicker rotor with more vanes dissipates more heat quicker than a smaller, lighter rotor. Maximizing rotor surface area is the key to heat dissipation.

When a rotor has more mass, it absorbs and dissipates more heat. More mass is created by a larger diameter, greater thickness and more vane area. Curved vane rotors have more vane area than straight vane rotors because the vanes are longer. Curved vane rotors create better cooling because of their greater surface area, and because of their shape, they move more air.

Having more cooling fins is not necessarily a weight disadvantage. If the vanes are designed thinly enough, there is little in weight gain. Going from a 32-fin to a 48-fin curved vane rotor of the same diameter and thickness increases the weight only 1.5 pounds (comparison is of Wilwood rotors). And, having more vanes adds to the strength of the rotor, reducing the chances of cracking.

The primary design parameter for any disc brake system is the ability of the rotors to dissipate heat. The temperature of the rotors and thereby the entire braking system is critical to maximum braking performance. If the rotors cannot dissipate heat quick enough, the brake fluid can boil and total system failure can result.

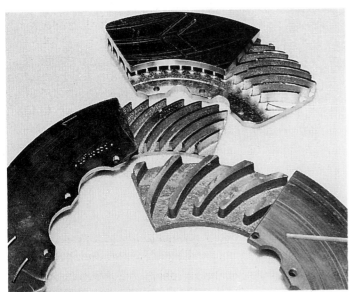

These rotor cross sections show (from top to bottom) a 48-fin rotor, a 32-fin rotor and a 24-fin rotor. Note the massive size of the fins in the bottom rotor. The bottom one weighs more, and yet has less efficiency, than the top one.

Super stick high coefficient of friction brake pads create higher rotor temperatures. This is because the driver brakes later and harder, and as rotor temperatures build up, the pads do not fade. With rotor temperatures rising higher, the rotor heat will increase and is going to stress the rotors as well as other braking system parts.

The higher heat build-up means increased stress. Rotors can respond with thermal shock, cracking and heat checking. When rotor surface temperatures increase rapidly, the rotor material expands. With high coefficient of friction pads, rotor surface temperatures can easily reach 1,200° to 1,600° F. When temperatures quickly rise and cool on rotors in short track cars, the rotors experience extensive thermal cycling. This can cause radial cracking of the pad contact surface from the corners of the rotor spoke supports, as well as heat checking on the rotor friction surfaces. The key to curing these problems is to dissipate the heat as quickly as possible. This is done through the selection of the proper rotors for the application.

Brake rotors store, then dissipate heat. The greater the metal mass of the rotor, the greater its storage capacity. If the heat dissipation or cooling is not adequate, the heat will travel into other areas of the braking system such as calipers, pistons seals and fluid, causing fluid boiling or piston seal failure.

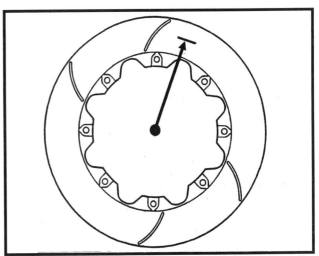

The effective radius of a rotor is the distance from the center of the rotor to the center of the piston contact area on the rotor.

Rotor thickness and diameter are both critical for obtaining proper heat dissipation and increasing overall braking effectiveness. And, because larger diameter rotors have greater surface area, they will dissipate heat better. Likewise, increased rotor thickness will absorb more heat and help dissipate it quicker. The rotor is what dissipates the heat and energy created by braking. If the rotor temperature exceeds the upper operating range of the brake pads, braking efficiency significantly decreases.

For a paved track braking system, a 1.25-inch thick rotor should be used with a diameter of 11.75 (or 12.19) inches in front and 11.75 inches for the rear brakes.

Drilling Rotors

Many racers drill holes in the rotor pad contact surface to remove weight, but this procedure causes many problems. These holes take away from the surface area and mass of the rotor, making it more difficult for the rotors to properly dissipate heat. The holes also cause a greater "thermal gradient differentiation," which means that the rotor will run much hotter on one side than on the other. This quickly leads to rotor cracking.

If a racer feels that he needs to cut weight from his rotors, it would be better to install a thinner or lighter weight rotor than to drill the rotors. He would end up with a much more stable and reliable rotor.

Rotor Mechanical Advantage

The effective radius of a rotor is the distance from the center of the rotor to the center of the piston

Good efficient air ducting for rotors is critical on short track cars.

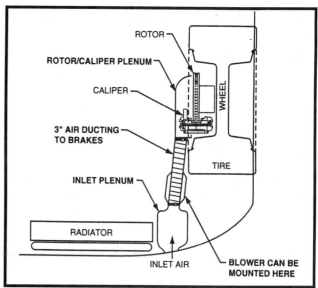

Air ducting plumbing diagram. Image courtesy of Wilwood Engineering.

contact area on the rotor. The effective radius is the lever arm that creates the work or stopping force against the rotors. The larger the rotor diameter, the greater the leverage exerted by the caliper on the rotating mass (axle or spindle). This translates into more stopping force. For good effective stopping power on short track cars, the largest rotor diameter possible should be used on the front wheels.

Bedding-In Rotors

All new rotors must be bedded-in to thermally condition them. Carefully following the proper bedding-in procedure will prolong rotor life, and make them more resistant to thermal checking and cracking. If new rotors are not broken in with the proper procedure, hard usage right away will cause thermal shock of the rotor which will reduce rotor life.

Before you start, make sure the rotors have all oil, grease and brake fluid cleaned off of them. To bed-in the rotors, run the car up to a moderate speed – one which allows the driver to make several hard braking applications without fully stopping the car. Make 5 to 6 moderately hard brake applications to slowly build up rotor heat. Then accelerate the car up to about 100 MPH (or maximum track speed, whichever is less) and make several hard stops to bring the car down to about 30 MPH. (While doing this, the driver should never drag the brakes as this causes radial hot spots on the rotors.) This should bring the rotors up to normal operating temperature (about

1,000°). Then take the car into the pits and let the rotors cool to ambient air temperature.

When bedding-in new rotors, used brake pads should always be used. Never use new pads with new rotors. Be sure to tape off cooling ducts when bedding-in rotors to help bring the operating temperature up quicker.

Rotor Air Ducting

When rotor temperatures approach or exceed 900°, good air ducting is critical. Ducting should run in as straight a path as possible to the center of the rotor using 3-inch diameter smooth interior wall hose. Intake plenums should be smooth and contoured to minimize air restriction. And use a course covering screen to prevent air flow restrictions.

Rotors act like an air pump. Centrifugal force pulls air from the center of the rotor through the vanes to carry away heat. In critical high heat situations, a second duct should be positioned to blow air across the calipers to help dissipate heat from them.

Rotor Maintenance

Extremely high operating temperatures cause the rotors to produce radial temperature variations. This means that the temperatures on the pad contact surfaces is not even from the top to the bottom. Rather, the peak temperature occurs in very narrow bands on the rotor face. This also means that these high radial temperature variations produce thermal stresses on the rotors, which can cause grooves and cracks.

Always check the rotors for signs of wear and cracking on a weekly basis. Any time a rotor has developed a crack, replace it immediately. As the rotor surface wears and develops deep grooves, the rotor has to be replaced. Rotors in race car applications can not be resurfaced like passenger car rotors. When the surface has worn, replace the rotor in order to maintain peak braking efficiency and safety. Heat checking, which is the appearance of small checked surface cracks, is normal and is not a cause to replace a rotor.

Whenever a rotor is replaced, always replace the mounting bolts with top quality new hardware. And always safety wire the bolts.

Check rotor run-out when rotors are new, and continue to check it on a weekly basis. Run-out should never exceed .006-inch. Top quality rotors are blanchard ground, so when new their surfaces are flat and parallel. If you find a run-out problem, many times it is a hub bearing or hub problem.

Brake Pads

Today's high technology brake pads provide high coefficient of friction operation over a wide operating temperature range, fade free operation at extremely elevated temperatures, and good wear resistance without excessive rotor wear.

Coefficient of friction (cf) is simply a measurement of the force required to drag one material across another, divided by the force holding them together. For example, if it takes a four-pound force to drag a ten-pound weight across a table, the the cf between the weight and the table is .4 (4 divided by 10). The higher the cf that a brake pad material has, the "stickier" or "more bite" it has.

The carbon semi-metallic pads used today are the highest technology in affordable braking for weekly short track racing. They combine the best attributes of semi-metallic and the aircraft industry's carbon/carbon technologies. This friction material produces a very high coefficient of friction without a high wear factor, but with a good consistency through a wide temperature range. And, the pricing is very affordable.

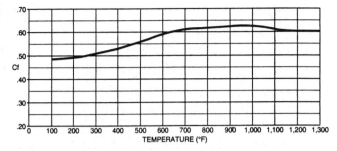

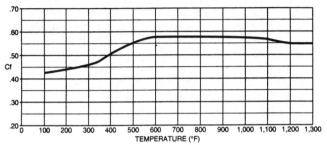

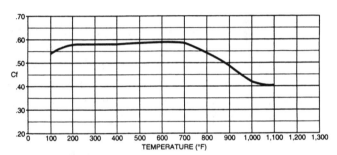

Note the operating temperature range of these three different pads. At top is the Wilwood Polymatrix B, center is Polymatrix C, and bottom is the Polymatrix D pad.

Some of the best carbon semi-metallic pads will reach their peak coefficient of friction at rotor temperatures of 1,000° F, and they will lose only .10 in coefficient of friction up to 1,600° F rotor temperature. A normal average coefficient of friction with this pad material is .35 to .42, with peak coefficient of friction at .50. These pads are formulated to operate efficiently at high temperatures.

Within the carbon semi-metallic group of pad materials, different compounds from different manufacturers will have an optimal temperature zone where the pad will work best. Each has a heat range where it operates most efficiently, and that is one of the major criteria for choosing the correct pad material. Brake pads are designed to work best over a certain designed temperature range. When they are operating below that temperature range, braking efficiency is reduced. When temperatures go above the designed temperature range, a drop in "cf" occurs and pad wear is acelerated.

Brake Pad Selection

Variables such as caliper piston size, rotor type and size, rotor operating temperature, race car weight, race length and driving habits all affect brake pad performance and choice. Other things to look at in selecting the proper pads are required braking coefficient of friction, consistency of use and pad wear and life.

Not all race tracks will require the same performance out of the braking system. Some tracks have wide sweeping banked turns that require a minimum of braking at turn entry. This type of track may only generate 600° to 700° of rotor temperature during braking, with good rotor heat dissipation down the straights.

Other tracks have long straights with tight flat turns that require very heavy brake application at turn entry. This type of situation may generate in excess of 1,000° to 1,200° of rotor temperature during braking, with less heat dissipation on the straights. So the same pad is not necessarily correct for all applications.

Consistency of operation and feel is also very important. The brake pad material has to feel the same to the driver across a wide temperature range so that the brakes operate consistently the same in the opening laps when temperatures are cooler through the end of the race when temperatures are extremely high. Consistency also means that the brakes can be used with the same type of feel and application during the life of the pad material.

Consult with brake pad manufacturers for a suggestion of the proper pad material for your particular application. They will have to consider the weight of your race car, how many laps you run, how heavy the brake application is on your type of track, how long between stops the rotors have to cool, and the type of calipers and rotors you use.

The ultimate way to choose a brake pad is by testing. Try different pads during an extended test session at your track and see which pad performs the best and which pad the driver is most comfortable with. Experience with different pads will also give meaningful feedback about pad wear and rotor wear.

Bedding-in Brake Pads

Disc brake pads should have a proper break-in process before they are used in a competition event. Bedding-in thermally conditions the pads, changing the molecular surface of the material. Proper bedding can double the life of a brake pad and increase the coefficient of friction.

Bedding-in new pads should be done by making a series of stops from moderate speeds, gradually building up the application rate until a slight amount of brake fade is felt. The brakes should then be allowed to cool for at least five minutes. The key to proper bedding is the process in which it is done. The pad material should build slowly from a low temperature up to a normal operating temperature. This procedure heat treats the pad friction material. If it is not followed properly, and the brakes are used hard with new pads, the friction material will be "cooked" and they will lose their friction properties.

When new brake pads are installed in the race car, always tape a note to the steering wheel so that the driver remembers to do the bedding-in process.

It is best not to bed-in new pads during practice before a race. The driver will not have the patience to do it correctly.

Some companies, such as Wilwood, offer a service for bedding-in brake pads and rotors, making both of these components ready to race. Wilwood uses a computerized dynamometer procedure to ensure that all pad sets are consistently bedded-in the same each time.

Never use new rotors when bedding-in new brake pads. Always have used rotors installed when bedding-in new pads.

Brake Fluid Recirculators

The purpose of a brake fluid recirculating system is to continually circulate the brake fluid throughout the entire system. This continuous circulation, actuated by the pumping action of the brake pedal, circulates the brake fluid through the calipers and back to the master cylinders. This eliminates localized heat build-up in the fluid in the calipers by continuously circulating cooler fluid through them. This helps to prevent fluid overheating or boiling in the calipers. The heat build-up encountered by the fluid is dissipated by radiation through the plumbing system as it is circulated back to the master cylinders.

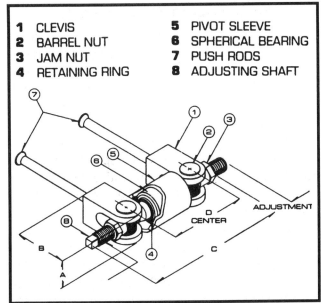

1	CLEVIS	**5**	PIVOT SLEEVE
2	BARREL NUT	**6**	SPHERICAL BEARING
3	JAM NUT	**7**	PUSH RODS
4	RETAINING RING	**8**	ADJUSTING SHAFT

This is the balance bar assembly from Tilton Engineering.

A brake fluid recirculation system should only be used when a race car is running longer than 50-lap events and brake fluid boiling is a problem. Recirculation systems require a lot of extra plumbing and they require regular maintenance. New brake fluid should always be flushed through the system prior to a race. Many times racers using a recirculation system become complacent about changing their fluid because they think the recirculation system keeps the fluid much cooler. It does, but the system still needs to be flushed before each race.

Adjustable Brake Proportioning

All race cars should have brakes with adjustable front-to-rear proportioning to adjust braking system efficiency. Paved track surfaces change according to track heat, weather conditions, oil or grease on the track, or the amount of rubber that is on the racing surface. All of these conditions will call for variation in the brake proportioning required between the front and rear brakes.

The purpose of a balance bar is to fine-tune the bias between the front and rear brakes. This allows the racer to achieve the correct ratio to maximize braking performance and handling. It is important to get into the ballpark with the correct piston sizes in the front and rear calipers first, but the fine tuning capability makes a balance bar mandatory to fine-tune the best possible brake balance front-to-rear.

Brake Balance Bars

A brake balance bar is a part which operates two master cylinders with one pedal. The pedal can apply force against an adjustable bar which inputs more force into one master cylinder than the other.

A brake balance bar can change brake proportioning to favor the front or the rear brakes. If a car pushes during braking/corner entry, add more rear brake bias. This restores more cornering traction to the front tires (refer to traction circle theory in the Tires chapter). Be sure to make brake bias changes in small increments — you don't want to make a big change and then have the car swap ends when you hit the brakes at the next corner.

The brake balance bar should be used as a fine tuning tool only. It should not be used to compensate for the use of the wrong size master cylinders. The balance bar is a great tool for fine tuning the chassis, and to compensate for changing track conditions. But it is not intended to compensate for large inequities in the braking system.

To calculate master cylinder input force with a balance bar, the total pedal effort must be divided by the proportion of the effort being fed into each master cylinder pushrod. For example, a pedal has a ratio of 6 to 1, and the driver exerts 100 pounds of force on it. That's a total of 600 pounds input force. If the balance bar bias is set 50/50 to the front and rear master cylinders, the force is 300 pounds (50 percent times 600 pounds) to the front, and the same to the rear master cylinder. If the balance bias is 60 percent front, 40 percent rear, the force at the front master cylinder pushrod is 360 pounds (600 times 60 percent) and 240 pounds to the rear (600 times 40 percent).

Balance Bar Adjustments

Master cylinder push rods should be adjusted so that when the pedal is retracted, both push rods are pulling against the retaining washers in the master cylinder. Any advancement of the push rod when the pedal is fully retracted will seal off the return bleed line and will lock the brakes on when the system heats up.

Master cylinder push rods should be adjusted so that when the brakes are on, the balance bar makes a right angle to the pedal. The balance bar should

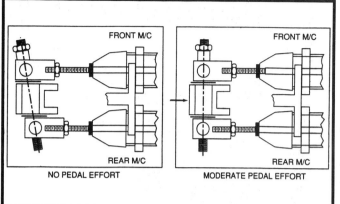

Adjust the master cylinder pushrods so that the adjusting shaft is parallel to the master cylinder mounting surface when the brake pedal is depressed. Adjust both pushrods an equal amount (one in, one out) so the pedal height is not changed. When the pedal is retracted, the adjusting shaft may not be parallel due to different sizes of front and rear caliper pistons. Image courtesy of Wilwood Engineering.

be parallel to the firewall of the car in the brakes-applied position, NOT in the fully retracted position.

A proper balance bar has a floating spherical bearing inside of a tube for its pivot and uses clevises with a barrel nut attachment to the push rod. Balance bars that use a spherical bearing fixed to the pedal and rod end bearing attached to the push rod are not proper as the balance bar will not stay in a horizontal plane when applied.

The length of the push rods in relation to each other have no bearing on the proportioning of the balance bar. The only effect the balance bar has on the leverage applied to the master cylinder is the relationship between the spherical bearing and the clevises. The clevis that is closest to the spherical bearing has the greatest force being applied to the master cylinder.

Brake Floaters

Brake floaters serve to remove braking torque forces away from other suspension movements and forces. They are separate brackets which are free to pivot about the rear axle housing, and are not welded to the axle tube. The brake caliper is bolted to the floater bracket, and a radius rod attaches the floater bracket to a chassis bracket. The radius rods direct the braking forces straight into the chassis and removes them from the suspension system. Adjusting the angle of the brake radius rods allows the

braking forces to be tuned to individual track conditions.

The radius rod angle can be set with the front angled up, straight ahead, or angled down. Each different angle creates a different effect on the tire at the corner of the car. For example, if the radius rod is angled up, the resultant force from braking adds force down on that tire contact patch. If the radius rod is angled down, the resultant force will lift up on the tire under braking. If the radius rod is set straight, there is no resultant force on the tire.

Adjusting The Floater Radius Rods

There is a brake floater and radius rod on each side of the rear end housing. Generally, the side which has the most upward angle of the radius rod will have the most downward force on the tire, creating more tire bite on that side. A good ballpark starting point is to set the left rear level and the right rear 10 to 15 degrees uphill.

Too much right rear uphill angle can make the car hard to turn in to a corner because the right rear is getting too much traction under braking. Too much

Brake floaters mount the brake calipers on a separate floating bracket, with braking forces input to the chassis through a radius rod.

left rear uphill angle can make a car loose under braking going into a turn, because the left rear will be getting too much traction and the rear of the car will want to pivot about that tire.

Start with the settings given above, then experiment with small incremental changes of the radius rod angles. When you have found an ideal set-up, measure the angles of both radius rods with an inclinometer, and record this in your notebook.

Braking System Check List

The calipers should be fitted to suitably strong brackets. The recommended thickness is 3/16-inch thick steel plate. The calipers must be square to the rotors in both planes, and must run in the middle of the pathway within plus or minus 0.015-inch. The bleed screw of the caliper should be positioned upward so as not to trap any air in the piston bores.

Check that all wheels clear the calipers. In cases where caliper interference is experienced, the entire mounting arrangement should be reassessed.

Curved vane rotors and differential bore calipers are single directional. They can only be mounted in one position. All of these parts have an arrow on them which indicates the correct mounting position and rotating direction. Make sure they are mounted correctly.

All brake lines should be secured to resist vibration, and the arrangement of the flexible hoses should be made so that interference with the wheel or tire does not take place at extreme steering lock, bump or droop.

The brake pedals should be rigid and braced for lateral load as well as beam strength. Under panic conditions, a driver can exert in excess of 300 pounds of force on a brake pedal. Any type of pedal mount flex will result in the driver's braking effort being used up in flexing the mount.

The fluid reservoir in the master cylinder should have sufficient capacity so that the master cylinder and fluid lines will not run dry when the pads have completely worn and the pistons are in full travel. At full pad wear, the reservoir should have at least a 1/2-inch level above the bottom.

The proper brake fluid for racing vehicles should be a glycol-based fluid which has a wet boiling point of at least 284 degrees F. For a racing vehicle that is properly maintained (which means frequent brake fluid changes to prevent the absorption of water into

The master cylinders and pedals have to be rigidly mounted or else the system will never develop full braking force.

the fluid), this is the only boiling point that is important.

Other Brake System Tips

Always match brake pad compounds on the front and rear brakes. Mismatching pad compounds front to rear can cause problems when the brakes are coming up to operating temperature, as their coefficient of friction will be different earlier than later. This can cause a very unstable braking situation. Another problem of using different pad compounds for the front and rear is that they can easily get mixed up during maintenance. It is difficult to identify which compound is which after the pads have been run.

Be sure that brackets which attach the calipers are installed absolutely parallel to the rotor running direction. Brackets which are out-of-parallel will cause a loss of braking torque because effort will be used by the caliper trying to square itself, and it will also cause excessive brake drag once the brake is released.

If you use a pedal arrangement that mounts the master cylinders on the floor, be sure to use a 2 PSI residual check valve in the lines from the master cylinders to prevent all of the fluid in the system from draining back into the master cylinders. However, if you have top-hung or firewall mounted master cylinders, do not use a residual check valve.

Master cylinders and pedals must be securely mounted on a body or chassis structure where pedal force cannot move the master cylinder mounting. If the mounting is not rigid, it has the same effect as having the brake lines expand.

Sorting Out A Braking Problem

Many times a problem of a continual push going into a turn on an asphalt track is not a chassis problem, but rather a brake proportioning problem. The front brakes are handling too much of the total percentage of braking effort, creating too large a slip angle at the front tires. Sometimes racers think of this as a spring problem, and stiffen the rear springs to balance the car. Then they end up with the classic problem of pushing going into a turn, and oversteer coming out. In reality, the problem can be cured by readjusting the brake proportioning (adding more rear brake bias).

If your car is pushing or loose consistently as you enter a turn, and you've tried various chassis adjustments to no avail, try this. Practice with the car at a reduced speed so that you can enter the turns without braking. If the chassis problem disappears, then you know the problem is in the brakes. For example, if the car was loose going in under braking, and now the problem has disappeared, you know that the rear brakes were working too hard in proportion to the fronts. Or if a push went away, your front brakes were handling too much of the total braking load.

Spongy Pedal Problems

Do you have a spongy brake pedal, and you can't seem to find the problem? There is a very simple, logical method to use in pinpointing the area of the system causing the problem.

Start at the right front wheel. Unhook the brake line from the caliper, and install a plug in the end of the hose. Now use the brake bleeding procedure we recommend, bleeding this corner of the car at the end of the hose instead of through the bleed screw of the caliper. What you have done is isolate the caliper from the rest of the braking system to see if the problem lies before the caliper or in it. If there is no air in the system before the caliper and you still experience a spongy pedal, you have air trapped in the caliper, or have caliper flex. (Be sure to bleed all four corners of the car with this procedure before coming to any conclusions.)

You can also use this same method to isolate a problem with the master cylinder, or in the brake lines. Unhook the brake line at the master cylinder, plug the line, then attach a bleed hose and bleed the master cylinder. If air is eliminated from the master cylinder and you still have problems, then you can probably assume that your problem is either air trapped in the lines somewhere or brake line expansion under pressure.

Braking System Maintenance

After every race, inspect the calipers, pads, rotors, hats, brake lines and master cylinders. On the calipers, look for cracks, fluid leaks, cracks or leaks at the bleed screw and fluid line connection, and cracks or problems at the mounting bolts. Check the pads for excessive or uneven wear. Check the rotors for cracking, galling, excessive wear or uneven wear. Check all lines for evidence of fluid leakage while they are under pressure, and also check all fittings. Check the master cylinder for evidence of leakage, and for proper fluid level. Periodically, the master cylinder should be disassembled and cleaned with brake fluid to ensure there are not any small particles of dirt which could clog the pressure relief hole.

Check the rotors often for cracks. If you find a crack that is anything more than a minor surface flaw, replace the rotor. Cracked rotors can lead to a rotor explosion on the race track, which is a very dangerous situation.

When changing brake pads, clean the exposed portion of the caliper pistons with brake cleaner. This will prevent pulling dirt and grit back into the piston seals and damaging them.

Change The Fluid Regularly

Most racers do not pay enough attention to the brake fluid. Fluid gets contaminated easily with moisture, lowering its wet boiling point. If a DOT 3 (284° F wet boiling point) brake fluid is being used, the fluid should be flushed out and replaced after every race night. Don't just add fresh fluid to the existing system, because you are mixing fresh fluid with contaminated fluid.

If you do not change brake fluid on a regular basis, use the Castrol SRF DOT 4 fluid. Although this fluid is much more expensive than DOT 3 fluids, it has a wet boiling point of 518° F, which is considerably higher than DOT 3 fluids. The expense of this fluid, changed once a month, will be closely equal to the cost of completely flushing the system weekly with a DOT 3 fluid.

Braking System Troubleshooting

Spongy Pedal

(1) Check to see if there is air in the system which is compressing. Bleed the system thoroughly.

(2) Fluid may be boiling because of water contamination. Drain the system, flush, and replace with new racing quality fluid (550-degree boiling point). Fluid will also boil because of overheated brakes caused by thin rotors, wrong pad material or brake drag.

(3) Check for leaks at bleed screws and line connections.

(4) Check for flex line ballooning under pressure. Have somebody apply heavy brake pressure on the pedal, and visually check all flex lines.

(5) Check for misaligned caliper. This can be spotted by tapered wear on pads, or if the caliper moves when the brake pedal is applied.

Pedal chatter, vibration or knocking

(1) This is primarily caused by rotor distortion. Check lateral run-out and caliper being mounted parallel with rotor.

(2) This could also be caused by worn suspension components, such as tie rod ends, ball joints or lower A-arm bushings.

Pedal fades as brakes are applied

(1) Check for fluid pressure leak at internal primary seal in master cylinder.

(2) Check for fluid leakage at line connections or along lines.

(3) Check for leakage at piston caliper seals.

(4) Too soft a friction pad material.

Low pedal, but pedal pumps up

(1) Low brake fluid level.

(2) Excessive free play in the pedal linkage.

(3) Pads are worn, causing excessive fluid use.

(4) Rotor lateral runout or parallelism problem.

Brakes drag or lock

(1) Master cylinders have a very small pressure relief hole, which, if clogged by a small piece of dirt, will cause the brakes to drag or lock. Check it. This hole can also be blocked if the master cylinder piston does not fully retract.

(2) The pedal return spring is weak or missing.

(3) There is insufficient pedal free play.

(4) Linkage bind prevents full return of pedal against its stop.

(5) Caliper piston seized.

(6) Improper caliper alignment to rotor.

(7) Distorted brake pad, or wrong brake pad being used.

Rapid brake pad wear

(1) Pad friction material is too soft or the wrong compound is being used for the rotor operating temperature.

(2) Rotor surface is rough or cracked.

Excessive pedal effort or excessive stopping distance

(1) The master cylinder bore size is too large.

(2) The pedal ratio is too low. Use at least a 6 to 1 pedal ratio.

(3) The pad material is too soft, or it has glazed over.

(4) Pad material has not been bedded in correctly.

(5) The mount of the master cylinder is not rigid enough, causing distortion.

(6) Brake pedal linkage is not rigid enough, causing distortion.

(7) Pads worn out.

(8) Unbalanced brake front to rear proportioning.

(9) Grease leaking on rotors.

(10) Wrong braking system for the car. Many times a racer will choose a light-duty system just to save weight — but it is not heavy duty enough for the application.

Chapter

6

Tires & Wheels

The vital link in any chassis set-up is tires. Tires, and more accurately, the tire contact patch or footprint, are the only connection a car has with the track. Racing tires have done more to improve lap times, as technology develops, than any other single factor. However, a basic understanding of what goes on is necessary to make sound choices about tire compounds, sizes, use, pressures, and how they affect handling. Let's look at some of the basic characteristics of the racing tire.

Coefficient of Friction

Any time two surfaces move in contact with each other, a coefficient of friction exists. The higher the coefficient of friction, the more friction there is present. In a case of a bearing within an engine, a low coefficient of friction is desirable. In the case of a racing tire operating on a track surface, the highest possible coefficient of friction is desirable. The higher the coefficient of friction, the more cornering force a tire is able to generate.

A number of factors affect the coefficient of friction of a tire. The construction techniques used in a tire itself include contact patch area, cord angle, tire compound, the section height, aspect ratio and other factors. Vehicle factors will also affect the coefficient of friction, the most notable being the vertical load upon the tire contact patch. The surface of the race track will also have a tremendous affect on the coefficient of friction.

The goal is to maximize tire contact with the racing surface and maximize cornering force. To do this, we want to maintain the highest possible degree of

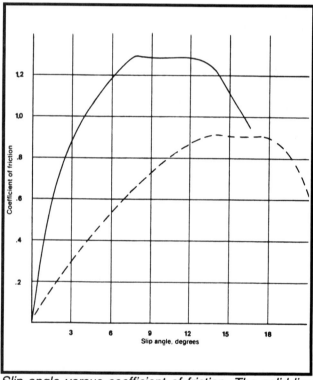

Slip angle versus coefficient of friction. The solid line represents a racing tire while the dotted line represents a bias ply pasenger car tire. Note that all tires will reach a maximum coefficient of friction, maintain it for a time, then drop off quickly when the slip angle becomes too large.

traction. Traction has two components – the coefficient of friction and adhesion.

Adhesion is the force which holds together the unlike molecules of substances when two surfaces are in contact with one another. Unlike friction, adhesion is a molecular interaction or interlocking. Friction is the resistance to the motion of two moving

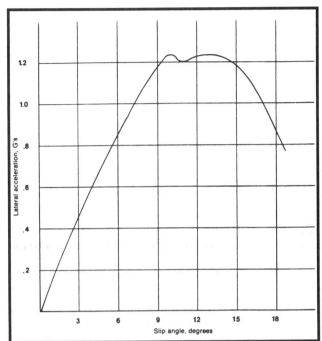

Slip angle versus lateral acceleration. The actual cornering force is related to the slip angle of the tire in a similar way as the coefficient of friction is related to the slip angle of the tire.

objects on surfaces that touch. For our purposes, the combination of friction and adhesion creates adhesive friction or traction.

The interlocking provided by adhesion allows a car to exceed previously determined limits in straightline acceleration, braking and lateral acceleration or cornering force. This is due to the ability of racing tires to achieve a coefficient of friction greater than one. A coefficient of friction greater than one means that resistance to motion between one surface and another is greater than the load applied to that surface. In this case, it is the tire surface against the track surface. Obviously, a change in tire compound or in track surface will alter the coefficient of friction as well as the ability of the tire to interlock with the surface of the race track.

Load

Load is the total weight that is on the tire. At rest, this is equal to the weight on each corner of the car. In the dynamic state, the load shifts due to lateral and longitudinal weight transfer. As the load on a given tire is increased, the total adhesion on that tire increases. Conversely, as the load increases, the coefficient of friction decreases. Adhesion is the amount of traction a tire is creating. When you add

load to a tire, you get more traction — up to a point. If you graph the curve of the increase, at lower G loading it is fairly linear (straight line). But as G forces increase, there is more of a curve and then a drop off. The heavier the car, the more rapid the drop off.

Loading versus cornering is a different story. A pair of tires (two front, or two rear) when evenly loaded, corner together better. The slip angle on a tire does not increase or decrease proportionally as vertical load is placed on the tires. And, more traction is available at a lower slip angle. So, during cornering, an outside tire loses traction quicker (because of more load from weight transfer) than the inside tire gains by a decreasing slip angle. The answer is to have a greater amount of static inside weight so during cornering the load on the left rear and right rear is closer. This helps to equalize slip angles, thus optimizing traction.

Tire Slip Angle

To make an oversimplification, the slip angle of a tire is the angular difference between the wheel direction and the contact patch direction. This has nothing to do with the steering angle. Both front and rear tires operate at a given slip angle any time a lateral load is generated.

Design characteristics of tires that affect slip angle are compound, cord angle, cord material, aspect ratio (a higher ratio equals a lower slip angle), and sidewall height versus tread width (a taller sidewall or narrower tread width increases the slip angle).

As the lateral load increases, the slip angle increases. The slip angle is also a determining factor of the coefficient of friction. As the slip angle rises from zero degrees, the coefficient of friction increases quickly up to a maximum figure between 8 and 12 degrees, depending upon tire characteristics. The curve flattens out at that point and gradually declines. It is at the peak of this curve

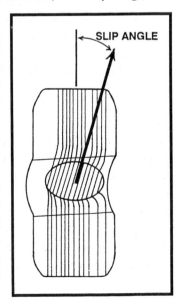

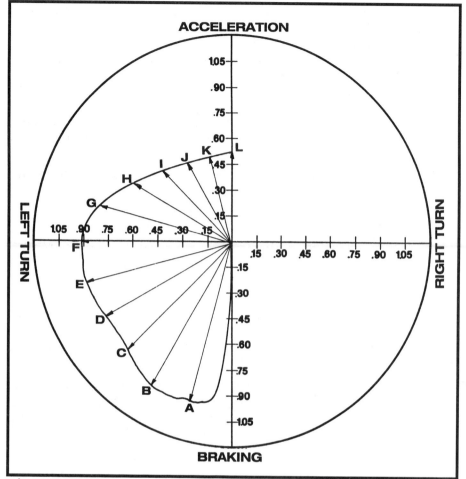

The arrows on this traction circle represent the various forces in G's being generated by the car in braking, cornering and accelerating. The solid line is a graphic representation of the traction capability of the car. At 'A', the car is braking. At 'B', both cornering and braking are taking place. At 'C', 'D' and 'E', there is progressively less slowing and more cornering. At 'F', all traction is being used for cornering. At 'F' through 'K', the car is accelerating more, with less traction being used for cornering. At 'L', all of the traction available is being used for acceleration.

where maximum traction is achieved.

Although an increase in vertical load will result in a net gain in cornering power, this will also increase the lateral load and can push the slip angle over the point where maximum coefficient of friction is achieved.

Keep in mind that the slip angle has nothing to do with the tire slipping or "sliding" on the surface of the track. When sliding does occur, it simply means that the curve of slip angle versus coefficient of friction has started to decline.

The *difference* between the front and rear slip angles determines the handling characteristics of the race car. If the slip angle is greater at the rear wheels during cornering, the car will oversteer. If the slip angle is greater at the front wheels during cornering, the car will understeer. The front-to-rear slip angle balance is adjusted by front and rear spring rates, or roll couple distribution.

A number of factors will affect tire slip angles. Spring rates will affect the loading on the tire and thereby the slip angle. Tire pressures will affect tire

slip angle, up to a point. Lower tire pressure will increase the slip angle at a given lateral force. Jacking weight will alter the tire slip angle by changing the loading.

The Traction Circle Theory

Tires really don't care in which direction they generate force — lateral, longitudinal, or any combination of the two. This fact has a tremendous impact on how a race car is set up and driven. The trick here is that a higher total force can be achieved by using the tire traction for two jobs at once, either cornering and braking, or cornering and acceleration. This phenomenon is represented by a circle graph called the traction circle.

The traction circle refers to the fact that traction is non-directional. Tires will provide a given amount of traction at the limit of adhesion. Tires don't care which direction traction is being applied as long as the limit is not exceeded.

If this is represented with a circle, the circumference of the circle is the limit of traction. This limit can be shown as a radius of the circle drawn in any direction. A radius to 12 o'clock would show 100 percent of the available traction used for acceleration. A radius to 6 o'clock would be 100 percent braking. The 3 and 9 o'clock positions show 100 percent lateral acceleration (cornering) to the right or left. A radius falling any place else on the circle shows a combined force in two directions.

In other words, 1:30 would show half the available force used for acceleration, half for cornering to the right. 7:00 would be a third of the force for cornering to the left and two thirds of the force for braking. There are an infinite number of points on the circumference of the traction circle. This whole concept dictates driving techniques.

The diagram shows a traction circle plot and relates it to the path of a race car through a turn. From point A to point E, both braking and cornering are taking place, with braking forces reducing as E is approached. At F, all traction is being used for cornering. From G through K, the car is accelerating, with less and less traction used for cornering. At L, all traction is used for acceleration. From here until the entry to the next turn, the graph will drop straight down the vertical axis as aerodynamic drag increases, horsepower falls off, and braking begins.

Keep in mind that on a traction circle diagram, any point on the plot not falling on the vertical or horizontal axis represents a resultant force with both a vertical and a horizontal component representing both lateral and straightline acceleration.

Braking Vs. The Tire Contact Patch

Utilizing the braking capability of a car is just as important as having a good braking system. A look at the traction circle theory shows that a tire has only a certain amount of traction capability in any direction. The maximum tractive force can be used up by accelerating, decelerating, cornering, or in a combination of any of the three. When hard braking is being experienced, the majority of the tire traction is used up by deceleration, leaving very little tire traction available for turning. What this means is a driver shouldn't be hard on the brakes while trying to turn into a corner. There just isn't enough tractive forces available to handle both situations. Braking should be done hard and quick in a relatively straight line before making a hard turn into a corner. The

driver should be almost completely off the brakes when he makes the hard turn into a corner.

Reading Tire Temperatures

Because all suspension adjustments and improvements are for the benefit of improving the bite of the tire on the track surface, it stands to reason that tire temperatures are the best indicator of what the tires and suspension system are doing. In fact, the tire temperature method is the only chassis tuning method where it is possible to get away from guessing and work with scientific accuracy.

The instrument that is used to measure tire temperatures is called a tire pyrometer. It is an electronic instrument which gives a temperature reading when its probe is inserted into a tire surface.

While you may think that a tire pyrometer is an expensive tool, it is not in comparison to the investment made in a race car. Prices for quality tire pyrometers are about the same as just one racing tire – or less.

Tire temperatures are read at three positions across the face of each tire and are recorded systematically on a sheet of paper in the manner shown in the accompanying examples.

By comparing the temperatures across the face of each front tire, it can be determined if each tire has too much or not enough negative camber, or if there is too much or too little toe-out, or if inflation pressure is correct.

Comparing the average temperature (of all three positions) of the right front tire to the average of the right rear will determine if the chassis is tending toward understeer or oversteer.

It is very critical to take your tire temperatures just as quickly as the car stops in your pit. The further

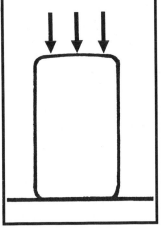

The tire pyrometer probe should be inserted across the tire face as indicated. The outside and inside edges should be measured about 1.5 inches inside the tire shoulder. Insert the probe at a 45-degree angle to the face, not straight in.

away from the track that you take the tire temperatures, the cooler the temperatures are going to be. And, the temperatures will be closer to each other across the face of the tire as they tend to even out because of conductive heat transfer.

Tire temperatures can change very quickly and cool off very fast. Take the tire temperatures very quickly, starting at the right front and working around the car in a circular motion to the right rear, ending up at the left front. Give the pyrometer needle time to warm up until the temperature stabilizes on the dial and then don't let it cool off when going from one tire to the next. This helps in getting a full set of temperatures quickly.

An easy way to prevent the pyrometer probe from cooling as you measure across tires is to keep your finger tip on the probe, then drag the probe across the tire surface to the next position. The probe will stay in contact with heat this way.

Always take tire temperatures in the same pattern. Start at the inside of the right front and work across to the outside, then go to the right rear, left rear and left front. Doing it the same way every time gives you reliable data for comparison purposes.

It always requires two people to take tire temperatures – one to insert the probe and shout out the numbers, and another to record the data. If a second crewmember is not available, have the driver record the data.

After you have checked the last tire on the car, take the temperatures of the right front again. This will give you an indication of how much the tires have cooled while you were taking the temperatures. This is very important for comparison purposes. What it may show you is that the left front tire isn't as cool as your data may indicate.

After tire temperatures are recorded, immediately check the tire pressures and record them. After tire temperatures have cooled, check the pressures again so you have an indication of pressure build-up due to heat.

Don't measure the surface tire temperature, but rather push the pyrometer needle (gently – they're brittle) in at a 45 degree angle approximately 1/8 to 1/4-inch into the tire. When inserting the pyrometer needle, you want to go through the rubber surface of the tire and touch the cords of the body. This gives you a truer and more accurate temperature of the tire. The inside and outside temperatures of the tire

should be taken about one-inch in from the shoulder of the tire (see diagram).

In the following discussions, the terms inside and outside edges of the tires are used. The inside edge *on both sides of the car* is always that one closest to the infield, and the outside is always closest to the grandstand.

Proper Tire Temperature Ranges

Most racing tires work optimally in a temperature range between 180 and 210 degrees. (The optimal operating range for harder spec type tires can fall between 150 and 175 degrees). Be sure you know what the optimum range is for the tires you are using. Ask your tire supplier or manufacturer.

If all of the tires are operating lower than the optimum heat range, there are three explanations:

1) The driver is not driving the car to the maximum traction capability of the tires.

2) The ambient air and track temperatures are cold, which cools the tires.

3) The tire compound is too hard for the application.

If all of the tires are operating much warmer than the optimum heat range, most likely the tire compound is to soft for the application.

Ambient Temperatures

When starting a testing session, be sure to measure the temperature of the air and the temperature of the track surface. If you start testing early in the morning, and continue for four or five hours, the air temperature and track temperature will get warmer. These factors will increase tire temperatures a proportionate amount. Continue to monitor air and track temperatures so you know and understand why tire temperatures continue to get warmer throughout the testing session.

Reading Camber

Perfect tire temperatures that reflect static negative camber settings should generally show the tire being at least 5 to 10 degrees cooler on the outside for a 1/4 to 3/8-mile track. The longer the track distance, the greater the difference should be between the outside and inside temperatures of the tire. The more static negative camber used, the greater the temperature difference. For example, if your car has 2 de-

---

Camber Reading Guidelines
Right Front Tire

Inside	Center	Outside
198	192	184

Indicates too much negative camber

Inside	Center	Outside
184	192	198

Indicates too much positive camber

Inside	Center	Outside
209	206	220

Right front is toed-in or too much positive camber -- depends on numbers on left front

Inside	Center	Outside
200	187	195

Needs more air pressure

Inside	Center	Outside
200	206	194

Too much air pressure

Inside	Center	Outside
202	196	191

Good setting for camber and air pressure

---

grees of static negative camber, there should be 6 to 8 degrees of temperature spread from the outside to the inside edge of the tire (outside cooler). At 1 to 1.5 degrees of static negative camber, the temperature spread should be between 4 and 6 degrees. At 3 to 3.5 degrees of static negative camber, the temperature spread is going to be 10 to 15 degrees warmer on the inside edge (these numbers will vary, of course, depending on the length of the track being run).

The difference in temperature spread with varying amounts of static negative camber is the result of the way the tire goes down the straightaway. With static negative camber, the outside edge of the tire is less heavily loaded. The more negative static camber, the less outside tire surface contact there is on the straightaway. This cools the outside edge of the tire.

The longer the track (or straightaways), the greater the difference between the outside and inside edge will be. On a short, tight track, like a 1/4-mile, the tire temperatures are going to be a lot more even across the face of the tire. This is because the corners are a greater percentage of the race track distance, and the outside edge has less time to cool on a real

short straightaway. So, the shorter the track, the more even the tire temperatures are going to be, outside to inside (with a given amount of static negative camber); the longer the track (especially longer straightaways) the higher the inside edge temperature should be.

Don't ever try to get the tire heat even across the face of a tire. If you do, you will wind up with a push problem going into the corner and in the middle of a turn. Getting the temperatures even across the face means the tire has too much positive camber, and the tire face will not be flat on the track surface through the corners. A bias ply tire corners best when it is cambered at –1°, so the right front inside temperature should always be warmer than the outside by at least 7° to 12°.

Average Tire Temperatures

A comparison of the average temperature of each tire can help determine how each corner of the chassis is working, which tire or end of the car is overloaded, and which tire is underperforming. To find the average temperature of a tire, add all three temperatures together, then divide by 3.

Beyond comparing averages of individual tires, it is also informative to compare the combined front tires average to the rear tires average, the right side average to the left side average, and the diagonal averages.

When one tire has an average temperature higher than all the others, it is overworked. There is too much weight on it – either static load or dynamic loading, or both. If the overheated tire is the right front, it may indicate that the driver is braking too hard and driving into a corner too hard. To check for this, have the driver change his corner entry and braking techniques, then re-evaluate the new temperatures. If the right front is still too hot, try a spring rate change – either softer at the right front, or stiffer at the right rear.

When the average temperature of one tire is much cooler than all the others, it is a good indication there is not enough weight on it – either static weight or dynamic weight or both. The most likely corner for this to happen is the left front. Even if the handling balance is fine and the car is quick, if one tire is much cooler than all the others, it indicates that tire is not handling its share of the work load. The car could be even faster if that tire developed more heat.

To get more heat into the left front, start with more static weight on that tire. The easiest way of doing that is readjusting the ballast, perhaps moving some ballast on the left side frame rail forward. Continue to readjust ballast until the tire temperature is reasonably warmer. After doing this, the spring rate at one or more corners may have to be adjusted to rebalance the car's handling.

Diagonal Tire Temperature Average

An analysis of the right front and left rear average temperature versus the average of the two front tires is an indicator of the correct amount of crossweight in the chassis. The right front/left rear diagonal average should never exceed the front average or the right side average. Ideally, it should be about 6 to 10 degrees lower than the front and right averages. If it is warmer, the car has too much crossweight and will tend to push. If the temperature is too low, there is not enough crossweight, and the car will tend to be loose because the left rear is not doing enough work.

Rear Tire Temperatures Versus Stagger

Stagger on the rear axle creates negative camber at the right rear, and positive camber at the left rear. So, with rear stagger present, you are going to see a temperature variation pattern on the rear tires. The inside edges will be hotter than the outside. The more stagger present, the more temperature variation.

Some Examples of Analyzing Tire Temperatures

The best way to understand the interpretation of tire temperatures is to read through some examples. Be sure to remember that references to the inside, in all cases, refer to the infield side of the car, and references to the outside refer to the grandstand side.

It is very important to understand that one overall look at just one set of tire temperatures will not give you all of the information at one time to correct the chassis. Sorting the chassis using tire temperatures is an evolutionary process. The testing should be done in ten-lap increments. It might require eight to ten tests to get all of the parameters adjusted.

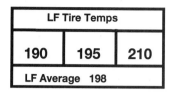

LF Tire Temps		
190	195	210
LF Average 198		

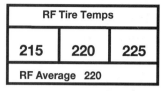

RF Tire Temps		
215	220	225
RF Average 220		

Left Average 194　　　**Front Average 209**　　　**Rt. Average 222**

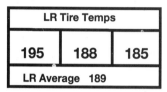

LR Tire Temps		
195	188	185
LR Average 189		

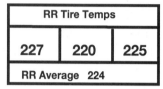

RR Tire Temps		
227	220	225
RR Average 224		

Rear Average 207

Example one

The adjustment parameters must be addressed in a specific order for the information to be totally usable: 1) camber, 2) tire pressure, 3) toe-out, 4) stagger, 5) crossweight, and 6) oversteer/understeer. This means that with your first set of testing temperatures, you should only analyze the camber for the front tires. Make the adjustments that the temperatures indicate, then test for another ten laps.

Continue this process with each parameter, in the proper sequence, until you have your chassis completely dialed-in. The reason that the tire temperature analysis and corrections are done in this order is so that one problem in one section will not mask problems in another area. Or, so that problems caused in one area will not be seen as problems from another area. For example, it is important to get the camber at both front tires squared away first before attempting to analyze toe-out. That is because a hot inside edge on the right front tire can be caused by too much negative camber, or by too much toe-out. One variable has to be sorted out and corrected before the other can be addressed.

Example One

The right front is warmer on the outside edge than the other two sections, indicating that it needs more static negative camber. The outside edge of the tire is warmer because it has been doing more of the tire's work than the rest of the tire surface, so it needs to be leaned in at the top.

The left front is too warm on its outside edge, so it needs more positive camber. That is, the top of the wheel needs to be leaned toward the inside (to the left as the driver sits in the car) so more of the tire's surface works on the track.

The right rear tire shows an average temperature of about four degrees warmer than the right front average. That would indicate this car is oversteering. Normally, an average temperature on the right rear of 10 to 15 degrees cooler than the right front would indicate a neutral steering condition (no oversteer or understeer). As the two average temperatures approach the same number, the car is on the verge of oversteer. When the right front is more than 15 degrees warmer than the right rear, the car is understeering.

Also on the right rear a slight case of underinflation of the tire can be noticed. If after several hard laps of driving the center temperature is warmer than the other two, the tire is overinflated. In this case, a difference of about five degrees cooler in the center indicates underinflation. If the center temperature difference was only 1 or 2 degrees cooler, it would not make any difference (remember that the maximum cornering power can be increased with the least amount of inflation pressure you can use).

The temperatures on the left rear, with the inside edge warmer, indicates a certain amount of stagger in the car. If this car was running on a 3/8-mile to 1/2-mile paved track and had 55 to 57 percent cross weight in the chassis, these rear tire temperatures would be too cool. Crossweight add more heat into the left rear tire. Adding crossweight into the chassis by making an adjustment at the left rear would help get more heat in the left rear and cool off the right rear temperatures, and take the oversteer out of the chassis.

Example Two

Looking across the right front temperatures, the outside edge is much cooler than the inside edge (the inside is almost 20 percent higher), so a little more static positive camber is needed to tilt the top of the wheel outward.

With the camber problem corrected, the new right front temperatures are 200, 190 and 185 (inside to outside), with an average of 192. Now the difference between the average right front and the average right rear temperatures is too similar, meaning that the car is on the verge of oversteer. Notice that the average left rear temperature is 25 degrees cooler than the

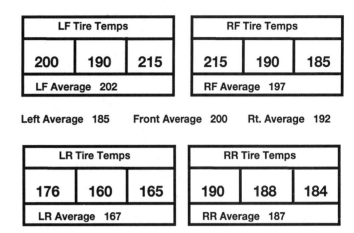

Left Average 185   Front Average 200   Rt. Average 192

Rear Average 177

Example two

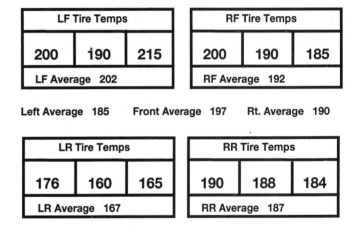

Left Average 185   Front Average 197   Rt. Average 190

Rear Average 177

Example two after right front camber correction

new right front average. That indicates the car does not have enough crossweight. Add more crossweight into the chassis at the left rear, and this will increase the heat at the left rear and right front tires, and decrease the right rear average temperature.

The left front, with the inside edge 15 degrees cooler than the outside edge, indicates it needs more static positive camber.

The left rear, with its cooler center temperature, could use a little more inflation pressure. The higher inside temperature indicates a normal stagger condition (there's only an 11 degrees spread from outside to inside – nothing to worry about).

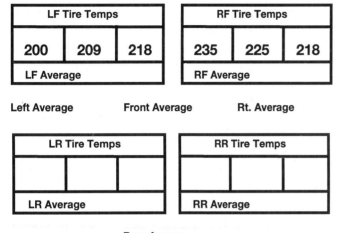

LF Tire Temps		
200	209	218
LF Average		

RF Tire Temps		
235	225	218
RF Average		

Left Average   Front Average   Rt. Average

LR Tire Temps		
LR Average		

RR Tire Temps		
RR Average		

Rear Average

Example three ---- too much toe out

The left rear has the coolest average temperature of any of the four tires on the car. If the cross weight was correct, the left rear temperatures would be warmer. The left rear needs some more weight on it to let it do more work.

Example Three

The toe-out indication is read on the inside edge of the right front and the outside edge of the left front. If the car has too much toe-out, the tires will heat up more in these two spots by the same amount of temperature excess. If there is too much toe-in, the outside edge of the right front and the inside edge of the left front will be too hot by about the same amount.

Example Four

This shows the ideal tire temperatures for a paved track car running on a 3/8-mile to 1/2-mile track, slight to medium banking. The average difference between the left front and right front tires is only about a 10 percent difference cooler for the left front. This indicates the left front is working. The heat pattern for the left front is slightly warmer on the outside. This shows there is more toe-out (Ackerman steering effect) at the left front, steering the inside turn radius more.

Left rear tire heat is good, showing enough crossweight to make that tire work properly.

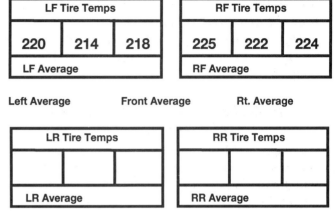

LF Tire Temps		
220	214	218
LF Average		

RF Tire Temps		
225	222	224
RF Average		

Left Average   Front Average   Rt. Average

LR Tire Temps		
LR Average		

RR Tire Temps		
RR Average		

Rear Average

Example three ---- too much toe in. If this car is running on a 3/8 to 1/2-mile track, the right front spread with ideal toe would be 225/220/215.

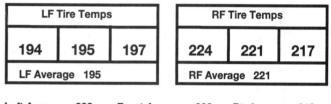

LF Tire Temps		
194	195	197
LF Average 195		

RF Tire Temps		
224	221	217
RF Average 221		

Left Average 203   Front Average 208   Rt. Average 213

LR Tire Temps		
209	204	199
LR Average 204		

RR Tire Temps		
210	205	200
RR Average 205		

Rear Average 205

Example four ---- ideal temperatures for a 3/8 to 1/2-mile track. Notice the average heat spread between the right front and right rear. Also notice the heat pattern across the right rear and left rear, indicating proper stagger.

The pattern of inside heat higher than outside on both the left rear and right rear shows the correct amount of stagger for proper rear tire "camber".

The patterns of heat going across the face of each tire from the inside to the outside, on all four tires, shows no drop or increase in the center position, which indicates there is no overinflation or underinflation of any tire.

The difference between the average of the right front and the average of the right rear shows almost a 10 percent greater heat in the right front, indicating a neutral handling car. The right front average is slightly more than 15 degrees warmer than the right rear, so the car is a little bit tight. The left rear/right front diagonal average is 213, which is 8 degrees cooler than the right front, and equal to the right side average, which indicates crossweight is right in the ballpark.

Problems That Cloud Tire Temperatures

Be sure that hard braking and cornering maneuvers from the last turn and getting off the track and into the pit do not influence the tire temperatures. If any of these have occurred, be sure to allow for it. Also, it takes a minimum of 10 to 15 hard laps of running before a tire warms up to proper operating temperature and any chassis evaluations can be made. The shorter the track and the cooler the track temperature (or the harder the tire compound), the longer it takes for the tires to heat up. Don't do anything to artificially induce heat into the tires to bring them up to operating temperatures quickly — it will only mask the true chassis characteristics that the temperature patterns should show.

And, as we said before, take the tire temperatures quickly. When the tires are hot right off the racing surface, the small subtle differences will show up. When the tires have had a few minutes to cool, they are lost.

Orderly Steps To Interpret Tire Temperatures

All of the information provided in the above examples can be very confusing. Reading tire temperatures and getting the proper and correct interpretation can be an art. It takes time and practice to determine what works. To help you in the beginning to sort out a set of tire temperatures, use the following set of priorities to determine chassis handling characteristics:

1) Look at all four tires first to see if there is any overinflation or underinflation present (center temperature significantly hotter or cooler). If so, adjust tire pressures, and take another set of test laps.

2) Look at the right front temperatures. Look for proper negative camber pattern.

3) Compare the right front average to the right rear average, looking for an indication of oversteer or understeer.

4) Compare the right front outside to the left front inside temperatures for an indication of toe-out or toe-in. Look for an indication of either too much toe-in or too little toe-out.

5) Look for the proper camber pattern at the left front.

6) Look for an indication of Ackerman steering at the left front (more toe-out at that corner only). The outside edge of the left front should be a little warmer.

7) Look for a proper stagger heat pattern at the left rear and right rear tires.

8) Look at the average temperature of the left rear and right rear, as well as the diagonal, as compared to the average of the front and the average of the right side for an indication of proper cross weight.

9) Look for all tire temperatures to be within the normal operating temperature range as prescribed for the tire you are using. Generally, the proper operating range for a short track tire is between 190 and 230 degrees. Check with your tire dealer or manufacturer for the correct temperature parameters for the tire you are using. If there is a corner of the car that is too hot, a spring rate change is indicated. For example, if the average of the right front is 260 degrees, and the right rear average is 210 degrees, the right front spring rate must be decreased or the right rear spring rate must be increased.

10) If the average of all four tire temperatures is either too high (above 230 degrees) or too low (below 190 degrees), you might take a look at a tire compound change. If the average temperature is too low, the compound should be softer (unless, of course, you are using a spec tire, which is harder and runs cooler). If the average temperature is too high, the compound should be harder.

11) While you are looking at the tire temperatures and making interpretations, do not ask the driver for his evaluations of the handling. Let the temperatures give you your first impressions, then compare that with what the driver has to say.

12) Make chassis adjustments, according to what the tire temperatures indicate, and have the driver make another set of test laps. Continue to do this until the temperatures show a well handling car and the driver is satisfied with the feel of the car.

Continue Monitoring Tire Temperatures

Camber requirements don't necessarily stay the same, even for the same track. So, tire temperatures should constantly be monitored. Track conditions and ambient air temperatures can change ultimate camber settings. And, if you pick up more speed, especially if you get into the turns harder, optimum camber settings will change. Continue to monitor tire temperatures.

Your starting settings depend on the camber change curve built into your car's suspension. If your car gains negative 1.75 degrees per inch of bump travel, and your buddy's identical car gains only negative 1 degree per inch bump travel, your initial camber setting at the right front is going to be less than his for the same track.

Advanced Tire Temperature Analysis

For more advanced tire temperature data analysis, take temperatures at various segments on the race track. Rent a track for a day so you can stop at any place on the track to take tire data.

Make ten laps on the track, then stop (not hard, not gradually) on the straightaway just past the second or fourth turn, and take tire temperatures. The collected data will show what the chassis is doing under acceleration from mid-corner and through corner exit.

If you want to know what the car is doing at corner entry, stop the car going into turn 2 or turn 4 and take temperatures. This will show what happens to the chassis under braking and turn entry. These track segment analyses may present results that differ from the average tire temperatures all the way around the track.

Reading Tire Surfaces

Reading racing tire surfaces is something of an art form. By reading the wear pattern on the surface of the tire, we are better able to determine handling characteristics of the vehicle. Combined with tire temperatures, sound judgments can be made about changes that need to be made to the suspension to improve handling characteristics. Often the ability to read the wear patterns is more valuable than reading tire temperatures. This is partly due to the

This tire came off the right front of a car. The outside is to the left, the inside to the right. The grain pattern across the center of the tire shows normal wear. But the outside edge (left) shows abnormal wear, and the inside edge (right) shows very light wear. These signs indicate the right front did not have enough negative camber.

characteristics of the race track where you are competing. As we have said, tire temperatures are more indicative of the last turn or two on a given race track. However, tire wear patterns will be indicative of what goes on all the way around a race track.

Normally, a tire's surface is a dull blackish gray color. If a somewhat shiny spot appears on the tire surface, this is normally an indication that portion of the tire is being overloaded at some point. If a portion of the tire surface shows no wear signs at all, this indicates the tire has no load or is very lightly loaded in that area. Because of the characteristics of racing tires and the desirability of dynamic camber change, the inside edge of the tire will generally wear somewhat more than the rest of the tire. If the wear is excessive, it indicates too much negative camber. If the outside edge of the tire is wearing more than the inside edge or the center of the tire, this is usually indicative of too much positive camber or not enough negative camber.

If little bits of rubber are rolling up on the tread surface, this indicates the tire is running too hot. If the grain pattern is too smooth, it normally indicates the tire is not running hot enough. The higher the tire compound temperature, the lower the tear strength of the tire surface. In other words, you can get the tire surface overheated to the point that track surface abrasion will pull chunks of the tire surface off.

If the front tires show more wear or excessive wear relative to the rear tires, this indicates the vehicle is understeering. If the rear tires show greater wear, this indicates the vehicle is oversteering.

If the center of any tire shows excessive wear, it indicates too much inflation pressure.

Developing the ability to read tires is important. One method which is effective to develop this quality is to study the tires of your top competitors. Compare the wear patterns and grain patterns on competitors' tires relative to their performance on the track. By practicing this, you will become adept at reading the tires on your own car and subsequently be able to improve handling and increase the performance of your own race car. Along with other records, you should maintain records of tire wear patterns and how they affect the vehicle on the race track.

Racing Tire Break-In Procedure

Racing tires are very influenced by heat the first time they are run. They go through a "final cure" in their first use heat cycle (first run on the race car). At this time the resins in the compound stabilize, the rubber cures, and the entire tire grows and assumes its final dimension. After this break-in final cure, the dimension of the tire is much more stable through the rest of its use. During the first heat cycle of the race tire, when it grows to its final dimension, it runs hotter than during its normal operating range.

Race tire break-in should involve gradually working the tire up to racing speed for a few laps until the tires reach normal operating temperature. They should not be run for more than a couple of laps at normal operating temperature, nor should they be abused or run hard the first time. If they are, they will change compound hardness or blister, and be ruined as a suitable racing tire.

The biggest mistake racers make is trying to get racing tires too hot too quickly. Racers use too low a tire pressure initially and ruin the tire. Tires should be broken in slowly, with proper inflation, running only two to three laps at racing speed, and then allowing them to cool.

Tire Use And Aging

When a tire is used over and over again, it gets harder and loses its grip of the track surface. The tire has been heated and cooled so many times that the compound oils are sweated out and the tire gets hard. At this point, it is very inefficient to grip the track properly.

Tire Stagger

Tire stagger simply means that one tire is smaller in circumference than the other at the opposite side of a solid axle. Tire stagger is generally used as a common term when the left side tire is smaller in circumference than the right side tire, and when the opposite is true, the term is called reverse stagger.

Stagger is used with a solid axle housing so that both tires will rotate at the same speed around a turn. Some cars require more stagger than others, with several factors determining this (including car rear track width, track turn diameter, tire construction, track banking angle, track surface condition, and most importantly, the type of rear end differential being used). The extremes of tire stagger on paved track stock cars commonly go from 0 to 5 inches. Tire stagger numbers are quoted as the difference between two tires' circumference.

The most important consideration in determining tire stagger is the type of rear end differential being used in the car. If the rear end is fully locked, both rear tires are locked together full time, and thus must rotate at the same speed. Tire stagger is very important here as a chassis tuning device to avoid understeer. The amount of stagger used is directly proportional to the radius of the track's turns. The tighter the turns, the more stagger required. The greater the turn radius, the less tire stagger needed. Generally, a minimum of 1.5 inches and a maximum of 4 inches of tire stagger is used with a fully locked rear end. Four inches of stagger would typically be used with a fully locked rear end on a 1/4-mile track, very flat, with very tight turns.

With a torque sensing differential in the rear end, the slip limiting device takes into account which tire is getting the best traction, and this allows the tires at the opposite ends of the axle to rotate at different speeds. With a torque sensing differential, the maximum tire stagger a car needs would be slightly less than required with a locker spool. Torque sensing differentials are also more forgiving of an improper amount of tire stagger.

Front Tire Stagger

There is no such thing as front stagger, because the front suspension is independent — the two front tires are not tied together on a common axle. Each tire is independent of each other and a circumference change at one will not affect the other. The only change that will be apparent is that a larger circumference tire is taller than a smaller one, so there will be a corner height difference, which is the same as weight jacking. Most racers, however, refer to this difference as tire stagger.

Once the front suspension is set for proper ride height and geometry, the last thing you want to do is use the front weight jack bolts to jack weight into the chassis. This will alter the front end and steering geometry. A better way to jack weight at the front is with front tire stagger. One-inch of front stagger is a common starting point. Front stagger adds wedge (or cross weight) to the chassis because the right front corner is higher than the left front. It puts tilt in the chassis at the front, which adds weight at the right front and left rear, and subtracts weight from the left front and right rear. Front stagger can be increased to add more wedge, or decreased to take some weight out of the left rear.

Tire Stagger Tips

The shorter the wheelbase of the car, the more critical that stagger becomes to handling. As the wheelbase gets shorter, the car becomes less forgiving of improper stagger.

High tire operating temperatures affect tire stagger more. If the handling is wrong and the tire temperature gets too high, the tire size will grow more quickly and tire stagger will get out of hand.

The most effective way to measure tire stagger is to use a caliper type of gauge. When using it to measure a brand new tire, be sure to take measurements in several places around the tire. This is because most new "sticker" tires aren't perfectly round, and numbers taken at different places can vary. It is best to scuff your tires first, which will make them round, before working with stagger.

Which Tires To Run?

To start out, you have three options: 1) Use the tires that your chassis builder recommends for your track or tracks; 2) Use what everybody else does at your

The most effective way to measure tire stagger is to use a caliper type of gauge.

track; or 3) Talk to your local race tire dealer or tech representative and follow his suggestions. Once you get the car working well with this combination, you may (you should!) want to experiment with other combinations, sizes, compounds, etc.

To minimize your guessing, you will want to: 1) Talk to a tire engineer from the company whose tires you have been running. Tell him what you have, what it's doing and what you want to achieve. 2) Next talk to an engineer from another tire company. Ask the same questions. 3) Talk to your chassis builder. He may be able to provide feedback from other competitors using his chassis or from his own testing and racing. 4) If you don't travel a lot, talk to racers who do.

Tire Air Pressure

Lowering the tire pressure will create a larger tire footprint on the track, but it will also create more tire heat. A change of 4 to 6 PSI in pressure can make a big difference. When the inflation pressure is too low, the tire will get hot very quickly. This occurs because the cords of the tire body have much more friction.

Proper tire pressures are a compromise between being low enough to provide good traction and grip, and high enough to support the shape of the tire so it does not distort laterally or roll under.

The correct tire pressure also allows the tire to use the entire tread face for traction, and prevents over-

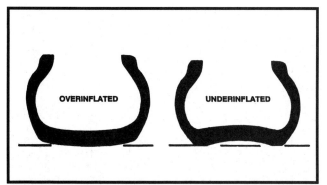

Overinflation causes the tire to be stiffer and reduces tire footprint on the track. Underinflation causes the tire to flex more, and reduces center tread contact with the track. Underinflated tires also run hotter because of the excessive flex.

working the tire which results in excessive heat build-up.

Tire pressures will vary according to the type of tire being used and track conditions. Ask your tire dealer for a recommendation for the particular tire, track, and car in question. And use a tire pyrometer to fine tune the pressure settings.

The total weight of the race car also has a bearing on the inflation pressures used. The lighter the car, the less pressure required. The heavier the car, the more needed. The reason is that with a heavier car, more pressure is needed to support the tire carcass and prevent deformation of it.

Because a tire is a flexible structure, it actually functions as a spring. And because air pressure is what supports that flexible structure, air pressure also determines the spring rate of the tire. The tire's spring rate adds to the spring's rate at each corner of the car to give an overall spring rate. So, adding or subtracting inflation pressure in the tire is going to add or subtract spring rate. This is a very critical element with radial tires, but it is also true with bias ply tires.

Adding more air pressure makes a tire's spring rate stiffer. Subtracting air pressure makes it softer. So, for example, when less air pressure is used in the right rear tire, it is the same as using a slightly softer right rear spring. And, what this does is tighten up the rear of the car.

Less air pressure creates more tire patch, or tire footprint, which in turn gives better tire grip. Less inflation pressure makes the tire slightly more flexible, which gives it a little more traction. But if a tire is too low on inflation pressure on a paved track, the center section of the tire will not be loaded properly and the tire will begin losing grip. The ultimate inflation pressure is a very narrow range. That is why it is so important to monitor inflation with tire temperatures.

Keep accurate records of how much air pressure builds up in each tire over a certain number of laps. Keep a record of track surface conditions, track temperature and ambient air temperature as well, because all of these factors influence how much air pressure builds up and how quickly. Generally, the majority of pressure build-up will occur in the first ten laps of racing. Lowering the starting pressures will allow you to control the pressure build-up in the tires.

Having accurate records and information will help you decide on the starting air pressures before each race. The goal is to end a race with the optimum air pressure in each tire.

Pressure Relief Valves

Tire pressure build-up is a problem to be aware of. One way to regulate the pressure build-up is to use a tire pressure relief valve. A pressure relief valve is a popoff valve which is mounted in the wheel and can be set for a maximum operating pressure. Once the tire reaches that pressure, the popoff valve opens and the excess air pressure is bled off. This valve can be very effective in chassis tuning because a deviation of as much as even 2 PSI from the optimum tire pressure can make a difference in tire performance.

Tire bleeder or pressure relief valves can be set to open at any chosen pressure setting. This allows a tire's pressure to be set at an optimum pressure, and the bleeder valve will maintain that pressure. This helps to control tire growth and thus helps to maintain the desired stagger.

In order to operate properly and prevent leakage, bleeder valves must be disassembled and cleaned on a weekly basis. Dirt contamination can lodge between the poppet and its seat, or clog the operating cylinder. Manufacturers recommend that the parts be cleaned with soapy water, blow them dry with pressurized air, and then lubricated with Armor All. Or, you can use Armor All to clean and lubricate the parts in one step. Don't use a petroleum solvent or part cleaning solvent because they will deteriorate the valve's seals. Be sure to put a cap on the bleeder

A tire bleeder valve can be set to open at any pressure setting, allowing the tire to maintain optimum pressure. This helps to control tire growth and maintain desired stagger.

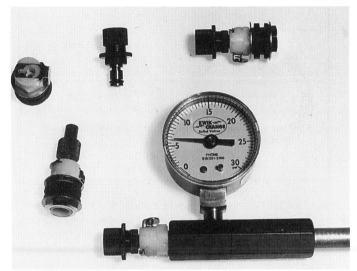

The Kwik Change relief valves are preset by plugging them into a small hand pump with an air pressure gauge attached to it. The knurled cap on the valve is screwed in or out to set the bleed-off pressure, then it is held inplace with the locking screw below it.

valve when you wash your car to prevent water and dirt from entering the valve.

To adjust the bleeder valve, first set the tire to the desired pressure. Then loosen the valve's adjusting cap until you hear the first hint of air escaping. Slightly tighten the cap until the air release stops. Then set the locking ring to secure the adjusting cap in place.

Aero Weld's Kwik Change Relief Valves

Kwik Change relief valves offer an innovation over traditional pressure relief valves. The Kwik Change bleeder valve can be taken out of the wheel to positively preset its bleed-off pressure. We say "positively preset" because to set the valve, you plug it into a small hand pump with an air pressure gauge attached to it.

To set the valve bleed-off pressure, plug it into the adaptor in the hand pump. (This adaptor is the same as the type as you plug into in the wheel.) Use the pump handle to build up air pressure. Adjust the cap on the valve up or down to adjust the bleed-off point. When it is found, use the locking screw to hold the cap in place. Pump the hand pump again to confirm the bleed-off pressure. Then plug the valve into the adaptor fitting in the wheel. The tire will be bled down to the preset pressure.

The plug-in of the valve to the wheel adaptor is a dry-break design, so it is instant with no air loss. The Kwik Change valve kit also includes a plug-in valve

stem to air up the tire if you don't want to use a separate valve stem. And, the kit also includes dry break solid plugs that go into the wheel adaptor when you wash the car to prevent dirt and soap film from entering it. There is also a release plug that can be inserted into the wheel adaptor to allow the tire to be deflated without removing the valve stem.

The unique thing about using this type of valve is how quickly pressures can be changed. Just take out the first valves and plug in the preset valves for lower or higher pressures. It takes about 30 seconds, and then you are free to do other work on the car. Be sure when you preset valves to a certain pressure that you use a marker to clearly mark the pressure on the valve.

Care should be taken to keep these valves very clean. Even very small pieces of dirt around the rubber seal of the valve can cause it to stick — either open or closed. The valves should always be thoroughly cleaned after each night of racing. Use a cleaner/lubricant such as Armor All on the valves. This will keep the rubber o-rings and seals clean and pliable without causing them to swell, and won't leave any sticky residue in the valve that will cause dirt particles to attract.

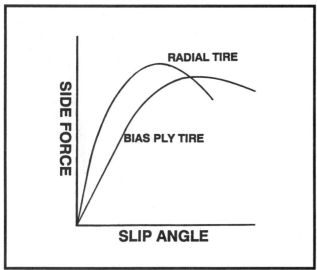

Both bias ply and radial tires operate at a slip angle in order to generate side force. However, the two types of tires operate differently. As you can see from the graph, a radial tire will build side force quicker than a bias ply. But the radial tire will drop off in performance much quicker once the peak side force has been reached.

Racing With Radial Tires

Racing radial tires, because of their style of construction, require changes in the chassis. Radial tires need a lot of initial negative camber setting (camber thrust) at initial turn-in to a corner. A camber change curve, even with a radical change, just isn't enough to get it done. You need the negative camber instantly. Winston Cup teams using radial tires are using in the neighborhood of 4.5 to 5 degrees of initial negative camber setting on flat to medium-banked tracks.

Radial tire sidewalls are very stiff. That changes everything. Soft, pliable bias ply tire sidewalls are forgiving, and add suspension and driver forgiveness. Radials do not. Radials require more driver sensitivity and smoothness.

A traction circle for a radial tire is different than with a bias ply tire. The circle of a radial tire is more of an ellipse, because the radial tire can generate more forward traction than lateral traction. This means that a driver has to change his cornering methods when using radial tires. The car should be driven into a corner straighter with radials. Then at the middle, the car is "rotated". Most of the turning is done at the middle of the corner so that the driver can drive out in a straighter line and thus apply more power earlier.

The one caution for a driver using this type of cornering style is that he should not get the car loose when he rotates it at the middle of the corner. This will happen if the driver applies power too early as he rotates the car. If it gets loose here, generally it will stay loose all the way through the corner exit.

Wheels

Before getting into any specifics about wheels for your race car, there is one absolute fact we must always remember about racing wheels: you don't want them to break under any circumstances. Wheels normally only break when under heavy loads — such as cornering — and that's exactly when you don't want anything to go wrong. Racing wheels seldom break under normal racing conditions. They do break when they are neglected or when inferior quality wheels are used to begin with. The most important element of wheel design is that they can withstand steering and cornering loads without flexing.

You should have all of your wheels Magnafluxed at least once every season or any time you suspect crash damage. Why? Although racing wheels are not cheap by any standard, they are a far sight cheaper than axles, rear ends, or chassis. And if you have wheels that fail, you can be certain that the following crash will destroy a lot more than the wheel. And it may destroy the driver. Using good judgment can save a lot of heartache.

Wheel weight and durability should be matched to the type of car they are attached to. A lightweight, thin-gauge wheel has no place on a 3,400-pound Street Stock car. Lightweight wheels are best used on much lighter weight cars. Heavy duty wheels are inappropriate for lighter weight cars because they will transfer crash impact damage to the suspension components.

Steel Wheels

Many wheel manufacturers offer a lightweight racing wheel which is made of a 0.075 to 0.100-inch thick steel rim shell. They are available in a wide variety of sizes.

Wheels are unsprung weight. Less wheel weight equals less unsprung weight, which is advantageous for better spring and shock absorber control over unsprung weight.

This is what happens to a wheel when you forget to properly torque the wheel stud bolts. ALWAYS check the torque before the car goes onto the race track.

In addition, wheels are rotating weight. The lighter the rotating weight, the quicker the car can accelerate and decelerate. It is a general rule of thumb that eliminating one pound of rotating weight is equivalent to eliminating four pounds of unsprung weight.

The lightweight wheel, because of it's thin gauge construction, will absorb impact better than the traditional steel wheel which is made of at least 0.135-inch thick steel. The wheels are designed to bend before destructive forces are transmitted to more expensive suspension and chassis components. This is accomplished with no loss of strength under normal racing conditions (such as not crashing). What this means is that in the event of a crash, they fold up and absorb the impact. This is opposed to the standard thick gauge steel wheel which lives through the crash but which transfers the forces to other parts such as A-arms or steering arms or radius rods or spherical rod bearings or housing tubes, which then are broken or bent.

Weigh the cost versus replacement frequency and repairability of both types of wheels before you buy.

Here are some tips for using steel wheels:

1) Steel wheels can be painted or powder coated. Don't use chrome plating.

2) Have steel wheels Magnafluxed at least once per season and whenever the car crashes or makes significant contact with anything.

3) Minor bends can be hammered out on the outer portion of the wheel.

4) Steel wheels can be welded. However, if a crash was severe enough to necessitate welding, have the wheel Magnafluxed after repairs are made. There may be stress cracks around the center section of the wheel.

5) Always inspect wheels for cracks around mounting lug holes.

Spun Vs. Rolled Steel Wheels

Comparing the way the material is formed in the two different processes, the spun wheel is stronger than the rolled rim. This is because the material can be controlled better in the spinning method. The corners and radii of the wheel are a more consistent thickness, which adds to the overall strength of the wheel.

The rolled wheel is formed by stretching the material over mandrels. When this is done, the corners and radii of the wheel are stretched thinner to conform. This weakens the wheel in these areas. The radii of the wheel must have a consistent thickness to keep the strength in this area of the wheel.

Wheel Balancing

Wheels should always be balanced, even for short track racing. Speeds can get up to the 70 MPH range even on the smaller oval tracks. An out-of-balance wheel is very annoying going down the straights. And while braking and cornering, the out-of-balance wheel will cause some handling difficulties. An out-of-balance wheel bounces up and down on the track, losing traction momentarily.

A badly out-of-balance wheel and tire can actually get off the track each time the wheel rotates, losing all traction for that moment. When the tire comes back on the ground, it will be overloaded and distorted, and it will again lose traction.

Wheel Trueness

All wheels should be checked for lateral and radial run-out, or trueness. Lateral run-out is sideways movement, commonly referred to as wobble. Radial run-out is the true roundness of the wheel. Both

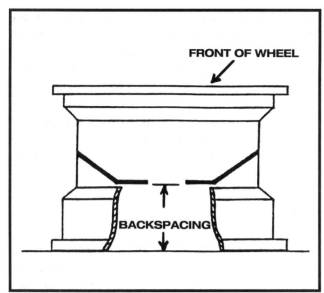

To measure wheel back spacing, lay the wheel with the back side down on the shop floor. Measure from the floor to the back side of the center section of the wheel.

dimensions should be checked with a dial indicator. The maximum allowable tolerance is 0.025 to 0.030-inch on both lateral and radial. If the wheel is new and exceeds these measurements, make the manufacturer exchange the wheel. If it's an old wheel, replace it. Check all of your wheels 3 or 4 times during the racing season for trueness.

Wheel Backspacing & Offset

What many racers refer to as wheel offset is really the backspacing of the wheel. Backspacing is the distance from the back face of the wheel to the back of the hub-mounting center of the wheel. This is the number that racers often refer to when they say "offset." The smaller the backspacing number, the further the center of the tire is located away from the centerline of the race car.

The offset of a wheel is the distance the hub-mounting center is located away from the true centerline of the wheel. So, for example, if you are using a 10-inch wide wheel with the center located on the centerline of the wheel, the offset is 0 and the backspacing is 5 inches.

Paved track stock cars normally use a backspacing ranging between 4 and 5.5 inches. The basic set-up for an "average" track is 5-inch backspacing wheels at all four corners of the car.

If a racer uses wheels that are too far outside of this range, problems will result. With too much back-

spacing, the tire/wheel assembly is going to start running into interference problems. With too little backspacing, the width of the wheel and tire exerts too much leverage over the hub and bearings, creating the possibility of failures.

Wheel backspacing can be used to help shift vehicle weight to the left side, as long as the amount of wheel backspacing is not limited by rules. A lower amount of backspacing on the right side coupled with a larger amount of backspacing on the left side shifts the car to the left within the same track width of the car.

Valve Stems

Use only steel valve stems and caps. Every time the tire is dismounted, check the valve stem for tightness. Always run with caps on the valve stems. Centrifugal force has been known to pull the valve stem open, allowing air to escape. Caps stop this.

Use the type of valve stem which has a cap that seals all the way down to the stem base. This protects the valve stem from impacts and debris. This type of cap also has a rubber gasket that seals against the stem base, which is an extra protection against leakage.

Regularly check for valve stem leaks. Use a spray bottle with soapy water and spray the stem and base. If there is a small leak, it will bubble out of the soapy water.

Tire/Wheel Preventive Maintenance

A good deterrent to tire leaks caused by debris penetration to the tire is the use of Marsh Racing Tires' Urethane Tire Sealant. The sealant is used to coat the inside of a racing tire, and when dry, it forms a tough air-tight film which seals air leaks. The stuff is a little expensive, but is very cheap insurance if you consider the alternatives.

New wheels should be tested for air leaks by installing a tire, inflating it and immersing the assembly in a dunk tank to look for leaks.

Bent wheels are a source of air leaks. If you suspect a leak with a possibly bent wheel, install a tire, inflate it, then paint the rim/tire joint with a liberal dose of 10% dishwashing liquid detergent/90%water, and look for bubbling.

Chapter

7

Chassis Set-Up

Proper chassis set-up is extremely important. You can't afford to have front tires that point the wrong direction in a turn, or a rear end that tracks at an angle — or anything that causes a drag on one or more tires on the car. Camber, toe and steering has to be right, and the rear end has to be square to the chassis as the body rolls in order to keep the tire contact patches from scrubbing off speed. Weight distribution has to be right to optimize cornering and acceleration. The car that has it all sorted out is going to have the advantage.

Checking For Chassis Binds

Before doing your chassis set-up in the shop, you must look for the presence of any chassis binds in your completed car. Move each of the wheels

Check for all possible chassis binds. The evidence here shows how much turning radius can be required. Note how the shock is in the way.

through at least two inches more than their normal wheel travel. Carefully observe the movement of everything attached to that wheel. Look for shock absorber binding or bottoming out, A-arms moving freely or contacting the frame, the steering shaft moving freely when turned without contacting anything, a free movement of all steering components through full range of left-to-right steering with no binding or contacting, the Panhard bar moving freely with no binds or without contacting any chassis parts, and the rear suspension arms moving freely with no binds.

If you observe any problems, be sure to correct them right away before proceeding to the chassis set-up.

Squaring The Rear End

Simply, the rear end squaring process is making sure the rear end housing is set straight in the car — perpendicular to the vehicle centerline and not angled. If the right rear is set behind the left rear, the car will be loose. If the right rear is set ahead of the left rear, it will push.

Squaring the rear end is very critical to the car's handling. Even a 0.25-inch out-of-square can have a significant effect on handling.

Most chassis manufactured by professional builders have built-in squaring reference marks on the frame rails. These references are usually holes drilled in the frame parallel to each other. Make sure you know where these are. If you have built your own chassis, make sure you remember to include these squaring references on the frame rails.

If your chassis is equipped with squaring reference marks on the frame rails, squaring the rear end is very

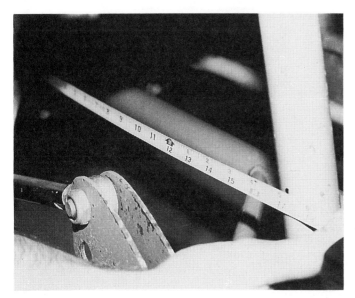

Most professionally manufactured chassis have squaring reference marks on the chassis that makes it very easy to measure from.

easy. Simply measure straight back to the rear end housing at the same point on each side from the reference marks. The measurements should be identical. If not, the rear suspension linkages must be adjusted to make them equal. If your car is not equipped with the rear squaring reference marks, you should add them. This is by far the quickest and easiest method of squaring the rear end. There will probably come a time when you will have to do it in a hurry at the race track, and you will be glad you added the marks.

Another way of measuring is to drop a plumb bob from the rear end housing to the floor, and then drop a plumb bob straight down from the frame reference mark. Measure between the marks on the floor on each side of the housing to make sure they are equal. When using this method, the frame must be level on each side.

The other way of squaring the rear end is commonly called "stringing" the car. It is done by stretching string tightly between jack stands on each side of the car running parallel to the wheelbase to establish outside reference lines. Measure from the string at several points to the frame rails to make sure the string reference line is absolutely parallel to the rails, and that each string is parallel to the other. Then measure from the string to the front side and the rear side of each rear tire (see drawing). Make sure the measurement hits the rear tire in the same point on each side. These measurements should be identical. If they are not, the rear end housing is not square.

If the front measurement on the right side is greater than the rear one, it indicates the rear end is pulled ahead on the right side. Adjust the suspension linkage which locates the rear end housing slightly, and take new measurements.

If a string is stretched between the two rear jack stands running parallel to the rear end housing, measurements can be made from it forward to the rear end housing on each side to double check the squareness, if there is room to measure. If not, drop a plumb bob to the floor from the housing tube on each side, then measure from the string to the plumb bob.

Once the rear end is squared, be sure the wheelbase is equal on each side. If linkages have to be changed to adjust it, make sure that the rear end is still square after doing so.

One of the most common problems that causes handling problems is a bent rear end. Many times a car has made contact and a

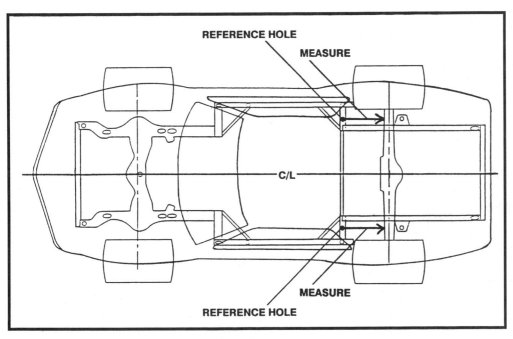

REFERENCE HOLE

MEASURE

C/L

MEASURE

REFERENCE HOLE

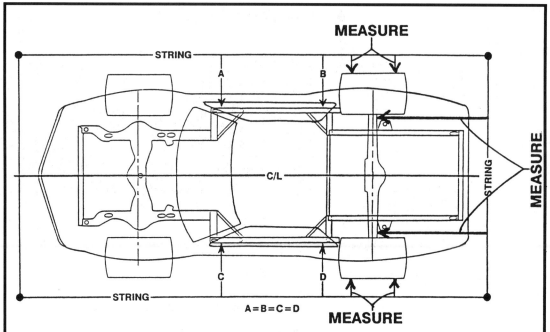

Stringing the chassis

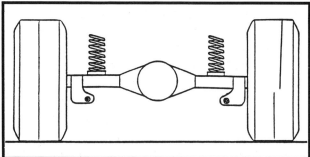

Checking for a bent rear end housing. Note on the right rear tire that the scribe marks don't line up.

The rear end housing must be centered in the chassis laterally. This is so that it doesn't create a torque steering effect. If the rear end is offset to the left or to the right, it could tighten the car up or loosen it during cornering through torque steer.

The lateral squaring is done by adjusting the length of the Panhard bar. Measure the distance from the back side of the brake rotor to the frame rail on each side of the car. Both measurements should be equal to square the rear end. If not, adjust the length of the bar to equalize them.

tube is bent slightly. To check for this, you need to take track width measurements. Put a scribe mark at the top of each rear tire in the center, and make a width measurement. Roll the car forward until the scribe marks move 90 degrees, then measure the track width again using these same marks. Continue doing this every 90 degrees. If the measurements differ, you have a bent housing tube.

Locating The Rear End Side To Side

While squaring the rear end, the Panhard bar must be disconnected. After rear end squaring, attach the Panhard bar and make sure that it locates the rear end housing centered in the chassis without offsetting it to one side or the other.

Setting Ride Height

The ride height should be set on the finished chassis in the shop before the weight distribution is set. Ride height is measured from the flats of the bottom of the frame rails at the most forward and rearward corners of each side of each rail. Make a mark on the frame rail so that you always measure from the same spot. Once ride height is set, the front/rear and left/right weight distribution can be set on scales with ballast. Using ballast to set the desired weight distribution will have very little effect on the corner heights.

For paved track cars, set the chassis as low as your track rules will allows (be sure the chassis is designed for the ride height you intend to use). Assuming the minimum ground clearance is 4 inches, use the following corner heights:

Left Front — 4"	**Right Front — 4.5"**
Left Rear — 5"	**Right Rear — 5.5"**

The tighter the radius of a turn, the larger the diameter of the outside tire must be in relationship to the inside.

Measure both sides of the car and record these measurements.

Rear Tire Stagger

Stagger is the difference in inches of the tire circumference between the left rear and right rear tires. When the right rear is larger in circumference than the left rear, we have stagger. When the left rear is larger in circumference than the right rear, it is called reverse stagger.

The minimum stagger required will vary from car to car and track to track. The variables include car track width (left rear tire center line to right rear tire center line) and the turn radius of the race track. Minimum stagger is the difference in size of the right rear tire from the left rear tire, determined by these variables. Because the two tires are running on different radii, one must travel further than the other (the outside tire must travel a further distance on a wider arc). This is accomplished by the outside tire being larger in diameter than the inside tire so it runs at a slightly faster speed.

The tighter the radius of a turn, the larger the diameter of the outside tire must be in relationship to the inside. So, the tighter the race track, the more stagger required.

Track conditions also dictate the amount of stagger used. If you are having a hard time turning the race car, stagger will help accomplish it. It will make the right rear overdrive the left rear and drive the car in a tighter arc — it makes the car turn itself.

Stagger helps get a car into a turn without pushing. And, it helps to overcome the pushing tendency of a car as it changes direction from straight line running.

Clearance blocks, such as seen to the right of the tape measure, will fit between reference surfaces of your chassis to help speed ride height setting.

Ride Height Setting Short Cut

Once your car's corner heights and corner weights are set, it is a good time to cut and fit clearance blocks which can help you quickly set up your chassis without the need for wheel scales, a flat surface and measurements at each corner. When each corner height and weight is finally set and right, there exists two relationships which will always remain true: the angle of the upper A-arms and the distance between the frame and the lower A-arm (at the front), and the distance between the bottom of the rear end housing and the frame at each corner of the rear – assuming that the chassis uses underslung frame rails at the rear. Knowing that these space relationships will always remain the same, small blocks of wood or metal (clearance blocks) can be cut which fit right between these reference surfaces. Once the car is perfectly set, fit and trim these blocks for each corner of the car, and record exact angle measurements for each of the upper A-arms. These can become your quick set-up tools whether you are at your shop or in the pits on an uneven surface and need to reset the car.

Another quick reference point on a coil-over car is to measure from the bottom coil-over mounting bolt to the center of the tube which is above the upper coil-over mount. Make a mark on the tube so it is easy to measure to and always repeatable.

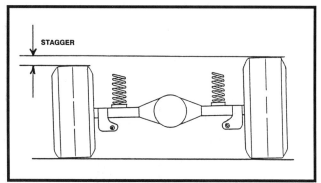

Stagger creates negative camber at the right rear, and positive camber at the left rear.

Tire stagger gets a car off the corners quicker, which means more straightaway speed. Improved corner speed with more straightaway speed helps you set up passes more easily. Does stagger hinder straight line speed? Yes — it's a dragging force. But more acceleration off a turn more than offsets this. You need enough stagger to launch the car off the turns to gain more straight line speed.

To find the minimum stagger your car needs, use the formula:

$$\frac{D + .5\,(TW)}{D - .5\,(TW)} \times C_L$$

where **D** is the track turn diameter in feet, **TW** is the rear track width of the car in feet (divide the inch measurement by 12), and **C$_L$** is the left rear tire circumference in inches.

For example, let's say a track's turn diameter is 200 feet, the car's rear track width is 64 inches (and divide by 12 to make it 5.33 feet for the formula), and the left rear tire circumference is 86 inches. Working those numbers through the formula would give you a minimum right rear tire circumference of 88.33 inches. So, the minimum stagger would be 2.33 inches. If your car uses a locking spool in the rear end, add 10 percent to that minimum stagger. If your car races on a track with 0 to 10 degrees of banking, or on a track with very tight turns, add another 10 percent. If you race on a track with 15 to 20 degrees of banking, subtract 10 percent.

From this you see that the tighter the turn, the more stagger the car needs. If a car pushes in the middle of the turn and beyond, more stagger is required. If the car is loose, decrease the stagger.

Street-type tires require more stagger, but it is harder to come by. By the design of the tire, street-type tires are not as sensitive to circumference change with air pressure increases. The best thing to do is measure different tires to find the stagger you want.

An important concept to remember with rear stagger is that it is also rear camber on a straight axle. If you put a smaller circumference tire on the left rear, it puts negative camber in the right rear and positive camber in the left rear. This helps put the tire footprint flat on the track during cornering.

Mass Placement

Mass placement is where major concentrations of weight are set in the car. Concentrated weight masses in the car include the engine, ballast, fuel cell, battery, etc. This placement is in relationship to the location of the roll centers, roll axis, and CGH. Mass placement interacting with these make a difference on how the chassis reacts with the roll of a car into a corner. Understanding mass placement is very important in understanding how the entire chassis works together because it goes hand-in-hand with roll centers and spring rates.

An important mass placement area to look at is the fuel cell. So many racers mount it in the chassis as far as possible inside the frame rails to the left. When the cell is full, before a race, that helps to achieve the desired left and rear weight distribution. But have you ever thought about what happens as fuel is burned off during a race? You are losing rear weight AND left side weight!

To see exactly what happens, we did a test with our **Racing Chassis Analysis** computer program. We defined our car as having a 108-inch wheelbase, 58 percent left and 51 percent rear weight, and using up 60 pounds of fuel (about 10 gallons of gasoline). The first car had the center of the fuel cell mounted 28 inches to the right of the center of the left rear tire. The second car had the center of the cell mounted 34 inches to the right of the center of the left rear tire. That only shifts the fuel cell 6 inches to the right. But the results were very interesting. When the first car (with the left mounted cell) burned 60 pounds of fuel, the left rear corner lost 26 pounds of total weight. When the second car (with the cell shifted 6 inches to the right) burned 60 pounds of fuel, the left rear corner only lost 23 pounds of total weight.

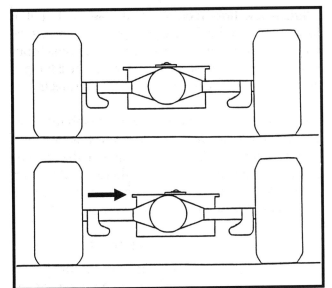

As fuel is consumed, the left weight percentage will decrease and the right percentage will increase. If the cell is located further to the right, the diminishing fuel capacity will not have as big an effect on the left rear corner weight.

This helps to explain why a race car's handling changes as fuel burns off. If the fuel cell is located close to the left rear tire, the left weight percentage will decrease and the right percentage will increase. If the cell is located further to the right, the diminishing fuel capacity will not have as big an effect on the left rear corner weight. And, generally as a race progresses, the track is going to get slicker and you are going to need more left rear weight to tighten up the car. You can help accomplish that with the location of the fuel cell.

The idea with mass placement is to get the weight placed in the chassis so that each tire contact patch is used to its ultimate during cornering. Build the chassis with all the components in place where they need to be, and keep the total vehicle weight in mind. Try to minimize total weight every place you can (but don't ever sacrifice safety). If everything is done correctly, you should have to add between 100 and 500 pounds of ballast to the car to bring it up to legal racing weight, depending on the minimum weight rule at your track. Then add ballast to the chassis in the places needed for proper weight distribution.

The ballast should be placed forward of the rear axle and between the frame rails. If you have ballast added behind the rear axle, it creates swing weight

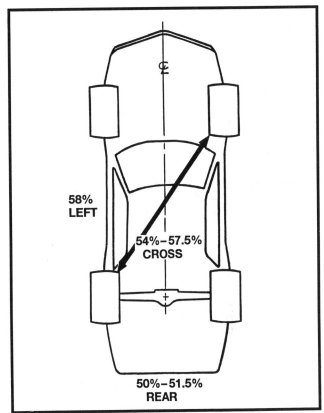

Suggested weight distribution

— like a pendulum — that will pull the rear of the car around as it enters a corner.

Read through the following "Chassis Set-Up At The Shop" section to see how ballast placement is done.

Suggested Weight Distribution

For paved track cars on all but high banked tracks, use the following weight distribution as a good starting "ballpark":

Left side — 58 percent
Rear — 50 to 51.5 percent
Cross weight — 54 to 57.5 percent

For left side weight percentage on a paved track, use as much as possible so that the left and right sides of the car are closer to being balanced during cornering. Usually, track rules will limit left side weight to 58 percent.

For rear weight percentage, 50 to 50.5 percent is used on faster tracks, while 51 to 51.5 percent is use on 1/4 to 3/8-mile tracks.

Crossweight is used to keep the chassis tight during cornering. It is advisable to run as little as possible. Use just enough to balance the car. More

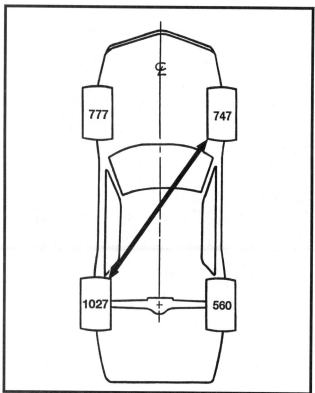

To find cross weight percentage, add together the right front and left rear weights, then divide by total vehicle weight. The numbers shown here equal a 57% cross weight.

crossweight heats up the right front and left rear tires more, making it a three-wheel race car. You want a balanced combination that gets more weight on the left front. The basic range of cross weight is 54 to 57.5 percent.

Cross Weight

Cross weight — or diagonal weight, as it is also called — is simply the total of the right front and left rear corner weights divided by the car's total weight. More cross weight adds more bite or understeer into a chassis. More cross weight is used to keep the rear end of the car tight on corner entry, and to improve bite off the corners. It favors the left rear tire contact patch by more heavily loading it.

To compute the cross weight, add together the single wheel weights of the left rear and right front corners. Then divide this number by the total vehicle weight. Your answer will be a percentage, which is the cross weight percentage. Cross weight relates to the same amount of percentage of weight set diagonally in the car, no matter what total weight it is, be it a 2,400 or a 3,200-pound car. You will want to

Many short track racers use a 58 to 60 percent cross weight, making a 3-wheel race car. This setup is fast in the early laps, but will fade in the last third of a race.

work with your car to find the diagonal weight that works best with it. Then, if you change the total weight of the car, you can still come back to that ideal bite set-up.

The total amount of cross weight required depends on track size and configuration, and the amount of roll couple used in the car. If the car uses a high front roll couple, the car is going to tend toward understeer. In this case, the cross weight is going to have to be a little less. If a lower front roll couple is used, the chassis will be looser, and more cross weight can be used to tighten the car up.

A car with a high horsepower/higher torque engine will usually use more cross weight, because the high power creates power oversteer. A lower horsepower car will require less cross weight because it doesn't have the power to break the rear tires loose and cause oversteer.

Many short track racers running a 30-lap Saturday night race use 58 to 60 percent cross weight. This makes the car a 3-wheel race car. It is fast in the early laps, but in the last third of the race this type of setup fades. It causes extra heat and wear at the right front and left rear tires. In order to have a strong setup at the end of the race, you must have a balanced setup that gets good grip from the left front and minimizes heat build-up and wear at the right front.

Cross weight works very closely with tire stagger. While cross weight helps put bite in the car for corner entry and exit, sometimes it will cause a car to push or understeer. At this point, more tire stagger is required to help balance the chassis. As a general rule of thumb, the more cross weight a car has, the more stagger it needs.

Chassis Set-Up In The Shop

Your goal in setting up the chassis at the shop is to have the car ready to race competitively as soon as it rolls off the trailer at the track. With the car set up properly at the shop, you should have to make a very minimum amount of adjustments at the track.

Be sure that you choose a flat, level surface in your shop on which to do the set-up. Always use the same place. Make marks on the floor where the car sets so it can be returned to the same location time after time.

The chassis set-up should follow a specific order each time. Make sure the correct springs are in the car, and that the car is completely race-ready. All fluids should be full, and the wheels and tires (including tire stagger) and air pressure should be the same as you plan to use at the track. Disconnect the anti-roll bar so that it will not hold weight in the chassis with preload. And, add the correct amount of ballast in the driver's seat to simulate the weight of the driver being on board. Only by taking all these steps can you guarantee a meaningful and consistent chassis set-up.

The order of chassis adjustment procedures should be:

1) Air the tires.
2) Set the stagger.
3) Set ride height at each corner.
4) Scale the car and adjust the weight.
5) Align the front end (camber, caster, toe).
6) Double check the weight distribution on the wheel scales.

Setting the proper tire pressures and stagger is the first step. They should be set to optimum racing conditions.

Put the car up on jack stands and air the tires to the desired pressures. At the same time, measure each tire circumference. Be sure to use a thin, flexible tape so there are no kinks or twists in the tape. Make sure you measure in the middle of the tire.

As you measure the air and circumference, mark your settings in chalk on each tire sidewall. Of course, when you're done with the chassis setup, rub off the numbers so your competition doesn't know what you're running.

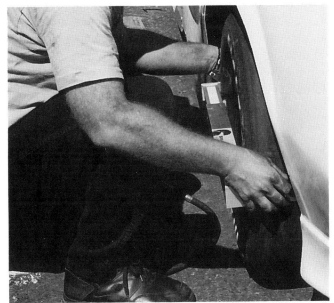

Start the chassis setup with properly airing the tires and then checking the stagger.

Setting the ride height is next, and that has been discussed previously in this chapter. However, for doing the initial weight distribution set-up, set the ride height at the right front and left rear corners 0.5-inch lower than the specified heights in the ride height section. This will help to get corner weights and cross weight correct along with the corner heights after the cross weight is set (this will become clearer later when we set the cross weight).

Scaling The Car

It is extremely important that all four wheel scales are set level to each other. If they aren't, the differences in height are going to add or subtract cross weight, or left or right weight in the chassis. Check that the floor is level from left to right, front to rear, and diagonally. Use a line level to check for accuracy. Just eyeballing it won't do. If a scale needs to be shimmed up, an ideal shim material is a 12x12 linoleum square.

Before electronic scales are used, they should be turned on (without any weight on them) and zeroed out. Read the instructions that came with the scales to learn how to do this. If you do not zero the scales first, you will get faulty wheel weight readings.

Wheel scales will help you get your car set up, but they must be used in the same manner every time, or you will not get repeatable results. When scaling the car to prepare the race setup, always make sure:

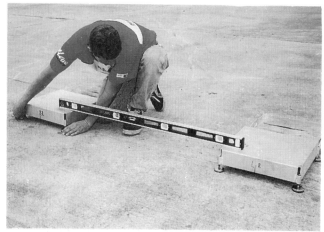

When scaling the car, it is important that all four wheel scales are set level to each other to avoid errors in corner weights.

Take the time to make sure each wheel scale pad is properly centered under the center of each tire.

1) The car weighs the same each time

2) The fuel load and fluids are always at the same level

3) The car is bounced up and down several times to eliminate any binding in shocks and springs

4) Tire pressures are always the same

5) Tire stagger is always the same, unless you purposely intend to change stagger to change your baseline setup

6) Each wheel scale is always centered under the center of each tire

The next step is to raise the car up in place and put a wheel scale under each wheel. Be sure to use a known quantity of weight, first, to calibrate the

When the car is scaled, the fuel load and all fluids should be full, and ballast that equals the driver's weight should be set in the seat.

scales and insure their accuracy. Once the car is in place on them, disconnect the shock absorbers (if the car has conventional coil springs) to be sure they do not hold the chassis up off the springs. Additionally, bounce the car up and down to settle the suspension. If the car has coil-overs, bounce the car up and down thoroughly to settle the coil-over units.

Read the wheel weights of your car and record them on a grid pattern on your paper for ease of visualizing the corners. Let's say that your finished race-ready car has to weigh 3,100 pounds with

A good proportion of the added ballast weight will have to go toward the left rear corner and the left side. With the car on scales, add ballast a little at a time and watch what happens to the corner weights.

driver on board, and you built it light, so the initial wheel weights are:

702	694
626	618

The total weight of the finished car is 2,640 pounds. The target weight of the race car ready to go racing is 3,110 pounds (the extra 10 pounds is added as a safety margin). So, 470 pounds of ballast will have to be added. The total left side weight is currently 1,328, or 50.3 percent. The total rear weight is 1,244 pounds or 47.1 percent. For a paved track application, the left side needs to be 58 percent, or 1,804 pounds, and the rear has to be 51 percent, or 1,586 pounds. So, it can be seen that a good proportion of the added ballast weight will have to go toward the left rear corner and the left side. Start by adding ballast to the chassis a little at a time, and watch what happens to the weights at each corner. Be sure to keep the ballast as low as possible, and make sure that all ballast, when finally set in place, is **securely** fastened to the chassis. Keep on adding ballast, and repositioning it, until the target wheel weights are achieved.

When scaling the car, be sure to keep a record of everything you do. Write down the corner weights and ride heights found at each corner when the car is first set on the scales. Then record every change you make and the resulting corner heights and corner weights.

Calculating Desired Corner Weights

The target corner weights for a 3,110-pound car on a paved track, with 58 percent left and 51 percent rear weight, would be computed as follows:

1. Use the target percentages to compute the percentages per wheel:
 A. Left rear wheel
   58% left x 51% rear =
   .58 x .51 = .296 or
   29.6 percent of the total weight on the left rear
 B. Right rear wheel
   42% right (100% - 58% left) x 51% rear =
   .42 x .51 = .214 or
   21.4% of total weight on right rear
 C. Left front wheel
   58% left x 49% front (100% - 51% rear) =
   .58 x .49 = .284 or
   28.4% of total weight on left front
 D. Right front wheel
   42% right x 49% front =
   .42 x .49 = .206 or
   20.6% of total weight on right front
2. Complete target weights:
 A. Left rear   = 29.6% x 3,110 = 920.6
 B. Right rear   = 21.4% x 3,110 = 665.5
 C. Left front   = 28.4% x 3,110 = 883.2
 D. Right front   = 20.6% x 3,110 = 640.7
Note: Round off all the hundredths of a pound — .50 or more to the next highest; .49 or less

(Left) Placing ballast affects left-to-right and front-to-rear weight percentages. (Right) Jacking weight from the weight jack screws affects the cross weight, but will not change the total left side percentage or front or rear percentage.

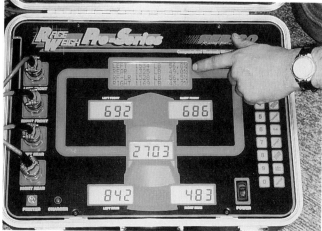

Many of the advanced models of electronic wheel scales will do all the math and percentages for you.

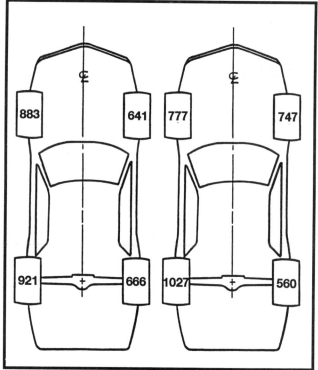

Attaching the ballast should help you achieve the target numbers as shown on the left. The desired cross weight, as shown at right, is achieved by adjusting each corner's weight jackers.

to the next lowest. So the correct target wheel weights would be:

Left rear	921
Right rear	666
Left front	883
Right front	641
Total	3,111*

*Add all four wheel weights together to double check your math. When we round up or down, sometimes it will add or subtract a pound from the total.

Many of the advanced models of electronic wheel scales will do all of the math for you automatically, but it is still important to know how it is done.

Obtaining The Target Cross Weight

If you add up the numbers, you will find that we have achieved the target 58 percent left weight and 51 percent rear weight, but the cross weight (left rear and right front added together and divided by the total weight) is 50.2 percent. This is lower than our target percentage, so some weight has to be added to the right front and left rear. The cross weight has to be set by screwing up or down on these corners' weight jackers. Screwing down adds weight, screwing up subtracts weight.

You will find that adding ballast will not adjust the cross weight. That will only change front-to-rear and left-to-right percentages. And, you will find that no matter what cross weight you use, the left side will always total the same and the rear will always total the same.

When the left rear and right front weight jackers are screwed down to set the desired cross weight, this will also raise the corner heights at these two corners. You should take this into consideration when setting the initial ride height of the car. Set the initial ride height 0.5-inch lower at these two corners.

To compute the desired corner weights which will yield the target cross weight:

1) Multiply the total weight of the car (3,110) by the desired cross weight percentage (57% or .57) which is 1,773.

2) Subtract from 1,773 the total of the cross weight we now have (921 plus 641 = 1,562), so 1,773 – 1,562 = 211. A positive number means we have to add weight to the left rear and right front. A negative number here would mean that we need to subtract weight from those corners.

3) Divide 211 by 2, which is 106 (rounded up).

4) 106 is the number of pounds we want to add to the right front and left rear corners by screwing DOWN equally on the weight jackers at those corners. If you have to change the corner height at the left rear and right front by more than 0.5-inch, screw

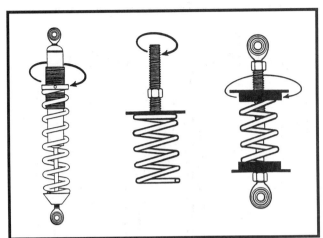

Whether you are working with a coil-over unit, a weight jacker over a coil spring, or a coil sping slider, the weight jacking adjustment method is all the same. On a coil-over, to jack weight in, or raise the corner height, turn the adjusting nut down. To take weight out of or to lower the corner height, turn the adjusting nut up. With a weight jacker, turn the jacking screw down to add weight, or up to take weight out. With the coil spring slider, adjust the top retainer plate the same as you would the adjusting nut on the coil-over.

While the race car is setting on the wheel scales, experiment with variables such as fuel load, stagger, and caster, that can change the corner weights of the car.

UP equally on the left front and right rear weight jackers until the proper cross weight is achieved. Adjusting all four corners will minimize changes in your chassis height adjustment.

After doing this, you will find that the added weight at the right front will have been subtracted from the left front, and the added weight at the left rear will have been subtracted from the right rear (assuming that your chassis is absolutely rigid).

The target corner weights for this car, with 58 percent left weight, 51 percent rear weight, and 57 percent cross weight will be:

777	744
1027	560

Be sure to add all four corner weights together again to make sure no math errors were made. And, double check the final cross weight math as well.

This same weight set-up procedure can be used on a car you have been racing for awhile if you want to start over and get a new chassis set-up baseline. Just follow the same procedure as we did above. Take all the ballast out of the car, and adjust the ride

heights at all four corners to the desired numbers. Then put the car on wheels scales and distribute the ballast and set the cross weight as discussed above.

Re-attach the anti-roll bar to the chassis while the car is still setting on wheel scales. That way you can see if the bar creates any type of preloading which increases or decreases cross weight. Ideally, the anti-roll bar will attach in a neutral condition – it will not add or subtract to cross weight.

Preloading the chassis with the anti-roll bar can add or subtract from the cross weight. Be sure to play with this adjustment with the car on wheel scales so that you understand what affect the preload has on cross weight, and how to achieve those results quickly at the race track.

Experimenting With Wheel Scales

While doing your setup with the car on the wheel scales, it is a good idea to experiment with variables that can change the cross weight as well as the left and rear weights. Those variables are fuel load, stagger and caster.

To check the affect of the fuel load on the weight percentages, start by putting the car on the scales with only 2 gallons of fuel in the cell. Calculate the left, rear and cross weight percentages and record them. Then add fuel. Check the weights at intervals

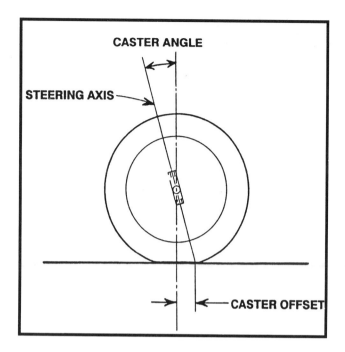

Radius plates are required to properly set the caster. Low profile plates like this from Reb-Co make setting the caster easy.

all the way up to a full fuel load. This will let you know how various fuel loads affect the chassis weights and thus the setup. Your ultimate goal is to find the best setup for your car at the end of the race when the fuel load is lightened.

Stagger affects ride heights, which will affect cross weight. For example, increasing stagger raises the right rear corner, and thus cross weight is reduced. Play with different stagger combinations on the wheel scales so you know how your setup is affected when stagger is changed. Record this information so you can make quick decisions at the race track.

Caster changes the ride height at the front corners of the car as the front wheels are steered left. More positive caster increases the ride height. While the car is on wheel scales, turn the wheels left and right to see how your caster settings affect cross weight.

Also record in your notebook what affect one round of weight jacked in at each corner of the car does to weight distribution. This information will be very important when you have to make quick adjustments at the track. You will know, from records and experience, how to make the proper changes to your chassis.

Front End Alignment

Caster

Setting the caster should be done before the camber because camber is affected by caster.

Caster provides directional steering stability. This influence is created with a line which is projected from the steering pivot axis down to the ground. This line strikes the ground in front of the tire contact patch when the caster is set in the positive position. A torque arm then exists (called caster offset) between the projected steering axis pivot line and the center of the tire contact patch. The torque arm serves to force the wheel in a straight ahead direction. The greater the length of this torque arm (caused by greater amounts of positive caster), the greater the steering effort required to turn the wheels away from their straight ahead direction.

The difference in the caster setting between the left front and right front is called caster split or caster stagger. A slight amount of caster split helps the car change from its straight ahead path more easily to ease into a left turn (when the caster at the right front is greater than the caster at the left front).

The amount of caster and caster split used on a race car is influenced by the use of power steering. Cars with power steering can use more positive caster and more caster split.

If your car uses a stock type front stub, you should be aware that a Camaro and Chevelle chassis with

RIGHT FRONT CORNER OF THE CAR

FORWARD

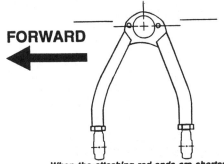

When the attaching rod ends are shortened equally, the upper A-arm is moved inward and the wheel gains negative camber.

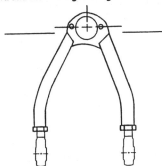

When the rod ends are lengthened equally, the A-arm is moved outward and the wheel gains positive camber.

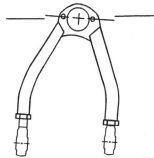

When the front rod end is lengthened, the top of the spindle is moved backward and positive caster is gained.

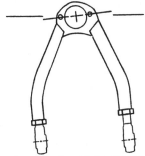

When the rear rod end is lengthened, the top of the spindle is moved forward and negative caster is gained.

How to adjust the camber and caster

the angled swing axis of the lower A-arms will develop a little more than .5 degrees of caster gain per inch of travel. The amount of caster gain also depends on the mounting angle of the upper A-arms in a car like a Chevelle or Camaro. If they are mounted parallel to the lower A-arm angle, the caster gain through bump is minimized. If the upper A-arm is mounted parallel to the frame rails, much more positive caster gain is experienced. Keep this in mind when choosing the caster settings for this type of car.

For a typical paved track chassis, use the following guidelines for the initial caster settings:

Manual Steering
Left front: +1°   Right front: +3°

Power Steering
Left front: +1°   Right front: +4°

To measure caster, first turn the wheel outward 20 degrees, level out the gauge and zero it. Then turn the wheel forward 20 degrees ahead of center. Level the gauge and read the caster setting. The front wheels have to be setting on radius plates to accurately set the caster.

Camber

The idea of the camber adjustment is to keep the tire contact patch flat on the track surface at the maximum point of cornering. Camber has the biggest influence over the vehicle's cornering ability than any other alignment feature. On most paved track cars, the static camber setting at the right front tire is between 3 and 3.5 degrees negative, depending on the camber change curve of the suspension, the type of track and banking, the tire construction, and the tire width. In general, wider tires use less initial camber, and narrower tires use more.

Use the following guidelines for the initial camber settings (track testing may show that initial settings may need to be changed):

Left front: +1.75°   Right front: −3.5°

Tire temperatures taken after practice laps will help you determine the exact camber requirements for your application.

Toe-Out

The amount of total toe-out that is correct for your application depends on the car's front track width, amount of Ackerman steer and the turn radius of the race track. The basic range of toe-out is 1/16 to 3/16-inch. The tighter the track and turn radius, the more toe-out required. Faster tracks with a larger radius and wide sweeping turns require less toe-out. Cars with a greater amount of Ackerman steering correction require less static toe-out, regardless of the track size, because it creates dynamic toe-out.

Start with an initial setting of 1/8-inch out, which is acceptable for most applications. If you have a fairly stock front suspension system, such as with a Street Stock, Hobby Stock, Pro Stock, etc., start with an initial setting of 1/4-inch out.

Tire temperatures and tire readings taken after some practice laps will help you determine the exact toe-out requirements for your application.

Setting The Toe-Out

The range of toe-out used on paved track cars varies from 1/16-inch to 3/16-inch. More toe-out causes the front end to stick more as the car enters a turn. If the car is sticking too much in the front, use less toe-out. Don't set the front toe-out at more than 3/16-inch. It would cause the front end to scrub off too much speed down the straightaways. Start with an initial setting of 1/8-inch for most applications.

Toe-out is set by measuring the difference between the front of the tires and the rear of the tires. There are several ways to accomplish this, but one of the easiest is to use a trammel bar set solidly against the right front tire. The other end of the bar has a pointer set a few inches away from the left front sidewall. Measure from the pointer on the bar to the sidewall at the same height front and rear on the tire.

When measuring to the sidewalls to find and set toe-out, you have to be very careful about sidewall distortions. Tires can have high and low variations that can easily make your measurements off by at least 1/4-inch. So before measuring for the toe-out, run through this procedure first to assure you have valid reference points:

Jack the wheel up and rotate the tire against a fixed reference point, such as a jack stand. In doing this you are looking for the highest and lowest spot on the sidewall. Mark the highest and lowest spots with

One of the best ways of measuring toe-out is with a trammel bar.

chalk, and then put those marks at the top and bottom when you are doing your measurements for toe-out. Every time you make an adjustment on the linkages to change the toe, you have to roll the car back and forth before you re-measure. With the marks at the top and bottom, you know you are getting the same reading every time.

Once an adjustment has been made, roll the car back and forth to take up any slack in the linkage. If the car had to be jacked up to make the change, also bounce the car up and down before taking a new toe measurement.

Double Check The Weight Settings

Once you're done with the alignment, check the car on the scales once again to make sure the alignment changes did not make any weight changes in the chassis. At that point, you're done and ready to go to the track! Make sure you don't bang the car around loading and unloading it from the trailer. That may affect the toe-out.

Some Hints For Working With Springs

Spring rates change, both in rate and height, after use. Variations are caused by the quality of the spring material and the type of use the spring receives. If you are unaware of the spring rate and height

When you get new springs, use a quality spring rate checker to double check the spring rate. Also, periodically take the springs out of the race car and check the rates to see if they have changed.

changes, it can adversely affect the handling of your car, and leave you wondering what went wrong.

When you first install new springs, use a quality spring rate checker to double check the rating (this is a good procedure to follow to guarantee that you got the spring rate you ordered), and measure the free height of each spring. Write these down in your chassis notebook, so you have a baseline to compare against in the future. Then periodically take the springs out and rate and measure them to see if there have been any changes. We have rated the springs on many race cars we have worked on, and have found that the actual spring rate is as much as 25 percent softer than the racer thought he had!

Starting Specs

The Recommended Starting Specifications charts show basic "ballpark" starting set-ups for an "average" paved track and three different types of cars. These numbers are based on a 3/8-mile oval, with up to 10 degrees corner banking. This set-up can work on a 1/4-mile and a 1/2-mile track as well, with probably a few adjustments. These set-ups are also a little on the "soft side". A softer set-up requires a smoother driving style and a more experienced driver. A lesser experienced driver might want a slightly stiffer chassis set-up until he gets more track time and learns the feel of different handling conditions. And, the inexperienced driver would do best with a car that has a little more push in the chassis — it won't be the fastest way through the turns, but it will be a stable feeling car.

Recommended Starting Specifications — Heavier Car

Track Type: Medium bank 3/8-mile
Car Weight: 2,800 – 3,200 pounds
Tires: Hard economy tires
Front Roll Center: 2 inches, alum. head engine
Rear Suspension: 3-point

Weight Distribution
58% left, 51% rear, 57% cross

Spring Rates (Conventional Coils)

LF - 1000	RF - 1050
220#/" anti-roll bar	
LR - 250	RR - 225

Spring Rates (Coil-Over)

LF - 350	RF - 350
220#/" anti-roll bar	
LR - 225	RR - 225

Shock Absorbers

LF - 76	RF - 76
LR - 95	RR - 95

Front End Alignment
Caster (manual steering)

LF +1º	RF +3º

Caster (power steering)

LF +1º	RF +4º

Camber

LF +1.75º	RF –3.5º

Toe-out: 0.125-inch

On a very short, very flat track, the right front spring rate will have to be increased to control excessive body roll.

Higher banked race tracks produce more downforce on the suspension, so all four spring rates have to be slightly stiffer to resist it.

Chassis Setup Variations

1) Front roll center/mass placement. The spring rates shown above are for a car with a 2-inch front roll center and an aluminum head engine. With a higher front roll center (2.5 inches) and a cast iron head engine, coil-over front spring rates would be 375 for the left front and 400 for the right front (coil-over springs).

2) Tires. These spring rates are specified for a car using a very hard, 11-inch wide economy tire. These spring rates will allow more body roll to get more downforce through weight transfer onto the outside tires. If softer tires are used, stiffer spring rates can be used, such as 400 at the left front and 425 at the right front.

3) Race length. The use of these spring rates assumes a normal main event length of 30 to 40 laps. If a longer racer —100 to 300 laps — is going to be run, it would be best to start the race with a stiffer right front spring. During a longer race, the track condition will get slicker as oil gets laid down, the tires get used up, and the chassis will get progressively looser. Using a right front spring that is 25 lbs./inch stiffer will make the chassis push in the early stage of the race, but as the race goes on and it is harder for the tires to grip the track, it will balance out and the car will be neutral handling.

If this type of spring setup is going to be used, be sure to qualify with the softer right front spring, then change to the stiffer right front spring for the main event.

4) Very short/very flat track. If you are running on a very short (1/4-mile) or very flat (0 to 6 degrees banking) track, the right front spring rate will have to be increased in order to control excessive body roll. If you are using a 350 lbs./inch spring at the right front on a 3/8 to 1/2-mile track with 10 degrees banking, when going to a flat 1/4-mile track use a 400 lbs/inch spring at the right front. A slightly stiffer right rear spring may be required to help balance the chassis.

5) High banked track. Higher banked race tracks turn the cornering force into more of a downforce than a purely lateral force. This means the spring rates have to be slightly stiffer to resist the downforce. Make the left front and right front springs 25 to 50 lbs./inch stiffer, and the left rear and right rear springs 15 to 25 lbs./inch stiffer.

6) Toe-out. While a moderately banked 3/8-mile track would require 1/8-inch toe-out, this would be changed for a fast 1/2-mile or a tight 1/4-mile track. On a fast 1/2-mile track, use 1/16-inch toe-out to

Big Bar Soft Spring (BBSS) Setups

An alternative to the traditional paved track setups presented here is the new big bar soft spring (BBSS) setup. This term refers to the soft front spring rates and very stiff front sway bar used. But there are many other changes required in the setup to make it work. Steve Smith Autosports has published a book on the BBSS setup, ***Paved Track Big Bar Soft Spring Setups***. See the ad on the last page of this book for more information and ordering details.

minimize tire scrub down the straightaways. On the tight 1/4-mile track, use 1/8 to 3/16-inch toe-out.

Recommended Starting Specifications — Lighter Car

Track Type: Medium banked 3/8-mile
Car Weight: 2,400 – 2,700 pounds
Rear Suspension: 3-point

Weight Distribution

58% left, 50.5% rear, 55 to 57% cross

Spring Rates (Conventional Coils)

LF - 950	RF - 975
220#/" anti-roll bar	
LR - 200	RR - 175

Spring Rates (Coil-Over)

LF - 300	RF - 325
220#/" anti-roll bar	
LR - 175	RR - 150

Shock Absorbers

LF - 76	RF - 76
LR - 946	RR - 94

Front End Alignment

Caster (manual steering)
LF +1º   RF +3º
Caster (power steering)
LF +1º   RF +4º
Camber
LF +1.75º   RF –3.5º
Toe-out: 0.125-inch

Recommended Starting Specifications — IMCA Modified

Track Type: Medium banked 3/8-mile
Car Weight: 2,500 pounds
Rear Suspension: 3-point, coils on sliders

Weight Distribution

58% left, 50.5% rear, 56.5% cross

Spring Rates

LF-650	RF-750
220#/" anti-roll bar	
LR-225	RR-200

Shock Absorbers

LF-76	RF-76
LR-946	RR-95

Front End Alignment

Caster (manual steering)
LF +.5º   RF +2.5º
Caster (power steering)
LF +2º   RF +5º
Camber
LF +1º   RF –2.5 to –3º
Toe-out: .125-inch

IMCA modified cars racing on paved tracks have a minimum weight of 2,550 pounds, as opposed to having no minimum weight for dirt tracks. An IMCA modified can be built with an average finished weight of 2,200 to 2,250 pounds. So, this means about 300 pounds of ballast can be added to the chassis which allows the racer achieve the optimum weight percentages. An IMCA modified paved track car can be built with more left side chassis offset to gain more left weight percentage. With carefully engineered chassis design and component placement, a 55 to 56 percent left weight percentage can be achieved before any ballast is added to the chassis. On a paved track, generally the cross weight will be about a percent or so less than the left weight percentage. An IMCA modified will generally use less cross weight than a faster late model sportsman because IMCA modifieds don't enter a turn at as high a speed.

Chassis Checks After The Race

Every week when you get the car back to the shop, check all the chassis settings before you disassemble

After every race, put the car back on the wheels scales to see if any weight settings have changed. Also check front end alignment and ride heights, and compare these numbers to your original settings.

the car or change everything. Be sure the car is setting on the same marks on the shop floor where you did the initial chassis set-up.

Check the weight at each corner of the car to see if anything has changed. Measure the ride height at each corner. And, check the front end alignment. Compare these numbers with your notes of how the car was set before the race. Then analyze any differences in chassis settings versus how the car handled in the race. The results can teach you more about your chassis.

Road Racing Setups

Whereas a chassis is heavily modified for turning only to the left on oval tracks, a road racing setup has to be "squared up" so the car can turn equally well to the right as well as the left. Several changes have to be made to an oval track chassis for this to occur:

1) Weight distribution. This is probably the most important change. The left and right side weight percentages should both be 50 percent when the car is turning both left and right. Both front and rear percentages should be 50 percent as well.

If the predominant number of turns are in one direction, such as mostly to the right, or if the fastest turns on the track are all in one direction, then the weight distribution can be slightly compromised to favor these turns. More weight – up to 2 percent – can be added to the inside to provide slightly better inside traction through the most important turns on the track. But, remember that the inside weight bias that helps the chassis turning in one direction will hurt the chassis turning the other direction.

Cross weight should be kept at 50 percent. If more crossweight is used to tighten the car up for left hand turns, this same weight bias will loosen the chassis up an equal amount in right hand turns.

There are several different theories about how to properly distribute the ballast to create a 50/50 front to rear and 50/50 left to right car. We think that the results we have seen in several winning stock cars on road courses favors the "perimeter weighting" theory. It is a lot like setting ballast for a dry slick dirt track. Weight is set higher in the chassis at the outside perimeters so that it quickly transfers to the opposite side to create side bite on the outside tires.

Place the ballast weight 15 to 18 inches above the bottom edge of the frame rails at the outside edges

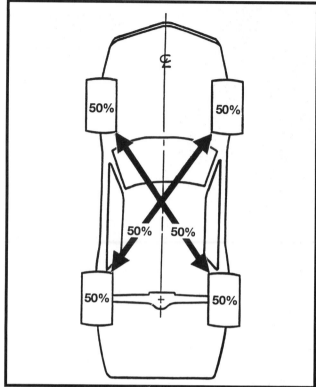

A road racing setup has to be "squared up" so the car can turn equally well to the right as well as the left.

of the frame. For front to rear weighting, place the front ballast at the firewall, and place whatever ballast that is required to equal 50 percent to 51.5 percent rear weight in the center of the chassis as far as possible toward the rear bumper.

The extra rear weight (beyond 50 percent) represents the amount of fuel weight that will be burned off between pit stops. So, the 51.5 percent rear weight would be set with the car having a full fuel cell. As the fuel is burned off during a race, the rear weight should never get lower than 50 percent.

To make this weight system work properly, a complete package of the proper shocks, spring rates and anti-roll bar rates must be used together.

The front shocks should be a 7 valve code (assuming a 2,800 to 3,200 pound car) from a manufacturer that makes their shocks much heavier on rebound control than compression (see the shock dyno graphs in the Shock Absorbers chapter). Heavy rebound control at all corners is a key to making this system work.

Rear shocks should be a 5 valve code, also of a type that has a much heavier control on rebound than compression.

On a road course, the anti-roll bar mounting linkages must be very solid and positive with no play. Mount the bar to the lower control arm within 1.5 inches of the ball joint center to get the most spring rate out of the bar.

For coil-over spring rates, they should basically be the same as used on a moderately banked 1/2-mile track. That would be in the range of 350 lbs./inch. However, the rates could be as stiff as 400 lbs./inch on the road course, depending on how high-speed the fastest turns on the track are. Left front and right front spring rates should be the same.

This type of weight distribution will create a lot of rapid overturning moment, and the springs alone will not supply adequate roll control. The front anti-roll bar plays a very important part in controlling the body roll. The anti-roll bar rate should be 100 to 200 lbs./inch stiffer than used on a 1/2-mile oval track. And – very importantly – the bar should be a *real* anti-roll bar. That is, one which is a machined bar with splined ends and machined bar stock arms.

The bar-to-control arm mounting linkages must be very solid and positive with no play. That is the only way the bar can properly control body roll during rapid transitions between left and right hand turns, and keep the car stable. The linkage has to be very positive with no slop or play.

The anti-roll bar must be mounted to the lower control arms within 1.5 inches behind the ball joint centers. This gets more spring rate out of the bar, and also keeps any play in the linkage from being magnified.

The heavy rate of the front anti-roll bar also plays another important part in the overall chassis design. It helps to hold up the inside wheel during cornering. Remember that both the left front and right front are set up for a negative camber change curve in bump. But when one of those corners is on the inside and

experiences rebound travel, the tire is inclined with a negative camber angle. With the heavier bar holding up the inside tire, excessive wear on the inside edge of the inside tire is prevented.

2) Shock absorbers. All four shock absorbers should have very heavy rebound control on road courses to keep the chassis tied down through a series of left and right hand turns.

3) Rear anti-roll bar. Consider using a rear anti-roll bar in the range of 60 lbs./inch on a road course chassis. It keeps the body flatter and more controlled through rapid body roll changes during left and right cornering.

4) Ackerman steering. Equalize Ackerman steering so it reacts the same when steering to the right as well as to the left.

5) Bump steer. Both the left front and right front should be set to bump steer .030 to .035-inch out per inch of bump travel. This bump out setting will create dynamic toe-out at the outside front tire no matter which direction the car is turning. The car needs this toe-out to come in right away at either side of the car as the car transitions from one turning direction to the other. Without the outside loaded tire toeing out, the car would push.

6) Spring rates. The spring rates for a road course would generally be the same at the right front and right rear as used by the car on a fairly flat to moderately banked 1/2-mile oval. The left front spring rate would be the same as the right front, and the left rear spring rate would be the same as the right rear on the road course.

7) Panhard bar. Use a full-width Panhard bar so that the rear roll center is located at the center of the chassis This puts the roll axis on the mechanical centerline of the car. This allows the car to roll equally left and right about one axis line. If the roll axis is offset, the car will not corner equally to the left and to the right.

The use of a short J-bar would cause roll center problems. With the J-bar attached at the center of the rear end housing and to the chassis on the right side, the rear roll center would be lowered through left hand turns, but rise going through right hand turns. The car would not handle the same through left and right turns.

8) Front roll center. The front roll center should be the same height as it would be for a short oval track, but it should be set on the mechanical centerline of

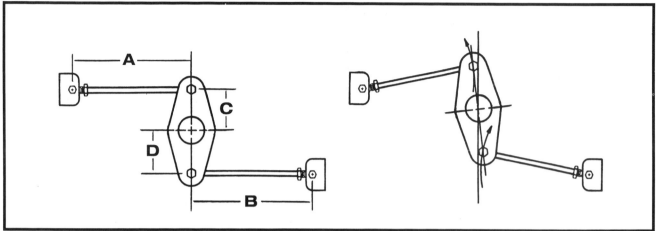

(Above and below) A Watt's link is an ideal lateral control linkage for a road course car when a quick change is being used. A Watt's link is the only type of lateral control linkage which results in zero roll steer. It keeps the chassis centered in its proper position over the axle during all movements of bump and rebound, without any roll steer resulting from a linkage arm moving the outside wheel through a forced arc. It is critical that links A and B be the exact same length, and that mouting lengths C and D on the center bracket be equal to each other.

the car. It should not be offset to one side. The front upper and lower control arms should be the same length side to side (uppers the same as each other, lowers the same as each other) in order to achieve this. The inner control arm attachment points of the left front should mirror the attachment points of the right front. The same spindle should be used on the left front and right front.

9) Camber curve. Adjust the camber curve of the left front to be the same as the right front on a flat oval. Both sides should be the same.

10) Camber and caster. Use the same setup specifications on the left front as you would on the right front on a flat oval track. Both sides should be the same. Don't use any caster stagger.

11) Tire stagger. Tire stagger cannot be used on a road racing chassis. Stagger is used on an oval track

to get the car to turn easier in one direction. But the use of stagger on a road course would make the car more difficult to turn in the other direction.

12) Rear differential. A Detroit Locker® or some type of torque sensing differential is a must on a road course. The use of a locking spool in the rear end would make the chassis understeer when turned both to the left and the right because tire stagger is not being used.

13) Brakes. Road courses are very hard on brakes. Be sure to use maximum ducting to the front brakes.

For maximum air ducting to the braking system, three hoses should be used, feeding into a rotor/caliper plenum mounted inside of the rotor.

Channeling large volumes of air from the front of the race car through the front brakes is required to remove the large amounts of heat generated by severe brake use on road courses.

On flat-front style of cars, the brake air ducting inlets should be mounted horizontally to make the most efficient use of the frontal air flow (see Figures 1 and 2).

Newer, more aerodynamically shaped vehicle fronts have changed the air flow pattern at the front of the car (see Figure 3). This diminishes the volume of air flow into horizontally mounted air duct inlets.

To make the most of the air flow patterns on a rounded-front car, the air duct inlets must be mounted vertically and as close as possible to the center of the race car (see Figure 4).

Air ducting hoses should run as straight and as be as short as possible in order to lessen any restrictions on air flow. Images courtesy of Wilwood Engineering

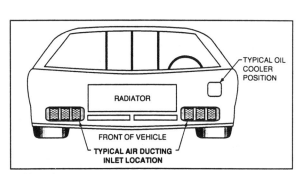

Figure 1.

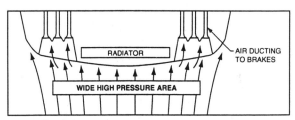

Figure 2.

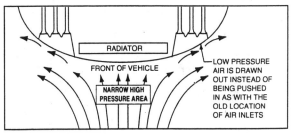

Figure 3.

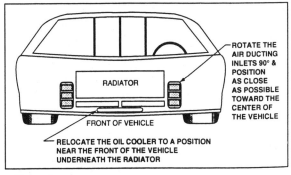

Figure 4.

Chapter

8

Track Tuning & Adjustment

Chassis Sorting at the Track

The car is set up properly at the shop. The springs and shocks should be correct. And yet, once the car hits the track, there are several handling problems to sort out. In this section, we will take you through these problems in case studies, isolating particular handling problems which are common to short track racers, and offer a diagnosis and correction for each.

When sorting the chassis at the race track, only one adjustment should be made at a time. If more than one change is made, you will never be able to determine the effect of just one change on the handling.

Always keep a notebook with you at the race track. Besides recording all the usual starting specifications of the car, you should also record each handling problem, and what adjustments were made to correct the problem. This log of information is important to you for two reasons: 1) Should the changes made not solve the problem, or should they make things worse, you know what changes not to make, and how to get back to your original setup. 2) It gives you a complete log of experience that tells you what changes on the car are effective for specific handling problems.

The first handling variables to get in line at the track would be the front end alignment. Tackle this first and get the tire contact patches operating at their maximum. Use a tire pyrometer to give you a positive feedback (see details on how to do this in the Tires chapter). And, use it to set the proper tire inflation pressure. Then check the shock travel indicators to see if any one particular corner of the car seems out of line (more on this procedure later in this chapter).

Defining Where A Handling Problem Occurs

When talking about handling through the turns, it is very important to isolate the three different phases of a turn. They are: 1) turn entry, 2) the middle of the turn, and 3) turn exit.

Each phase of the turn will be affected by what happens at the previous one, so it is important to understand this relationship. For example, one of the most common problems is a car that pushes going into a turn and is loose at turn exit.

What's happening is that the car is basically pushing. As the driver turns in, the car pushes. As he approaches the middle phase he is making corrections to loosen the car up. Then when he gets back on the throttle for turn exit, he adds to the oversteer problem and the car is loose on turn exit.

So where do you start to correct the problem? At turn entry. You have to get the push out of the chassis at phase one in order to get the car neutral handling

To correct a handling problem, you first have to define where the problem is occuring.

Phase one of a turn

Phase two of a turn

at phase two or three. Always cure the phase one problem first before working on anything else. Many times the curing of a turn entry problem will help to cure any other problems the car may experience.

Phase One Handling Problems
Oversteer

Let's say your car is oversteering, or loose at turn entry. What are the causes and what changes can you make to help correct that?

(1) There is too much rear brake bias — if oversteer occurs while braking into turn. To fix this, reduce the rear brake bias.

(2) The rear roll center may be too high. Lower the rear roll center by adjusting the Panhard bar height.

(3) The right front spring rate is too soft — add a stiffer spring.

(4) The right rear spring is too stiff — soften it up.

(5) The front sway bar is too soft — use a stiffer one or change to a different mounting position on the lower control arm to decrease the leverage ratio (see the Front Suspension chapter for more information on this). Make sure the bar is neutralized — no preload in it. If the bar is mounted to the frame with a sliding clamp mount, the bar can be slid back toward the lower control arms to shorten the arm length, which increases the bar spring rate.

(6) There is too much stagger. Decrease the stagger.

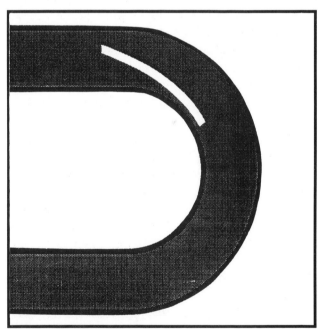

Phase three of a turn

(7) You need to add more cross weight — cross weight tightens up the chassis. Add weight at the left rear corner, and/or take weight out at the right rear corner.

(8) The right rear tire is over-inflated — decrease tire pressure. Take tire temperatures to confirm this.

(9) There is rear roll steer present in the chassis. That is, the left rear is moving ahead of the right rear during body roll, which steers the rear end of the car

If the car oversteers at corner entry, there are 9 different adjustments to look at. The easiest adjustments are brake bias, stagger, cross weight or spring rates.

If a car is pushing at turn entry, look at right front camber and too stiff of a shock at the right front as probable causes.

outward toward the wall. The oversteer is proportional to amount of body roll.

Use tire temperatures to help you sort out what the chassis is doing. Analyzing the temperature patterns should give you major clues as to where to adjust first. Then start with the easier adjustments first, such as changing rear roll center height, adding cross weight or decreasing stagger.

NOTE: If the car was pushing instead of loose at turn entry, we might also look at right front camber and too stiff of a shock at the right front. In the case of camber, if the tire does not stay upright as the car rolls into the turn, not all of the footprint of the tire surface is going to be on the track. The car is going to push, because all of the available traction surface of the tire is not gripping the track. If the camber curve at the right front is slow to come in, you might experience a push at initial turn-in, and then good traction at the front will return when the tire is located perpendicular to the track.

In the case of the shock absorber at the right front, you have to remember that the shock does not make any permanent changes to the chassis. But its rate of control can influence how quickly something happens. If the right front shock valving is too stiff, it can hold that corner of the car up at turn entry and cause the front end to push.

Tuning with shock absorbers is something that should be done after the handling is adjusted and set in all the other areas. Shock absorber damping characteristics are only used to fine-tune the handling of the car. They will not cure big problems, but they will cause a problem if they are too soft or too stiff.

If a car is loose in the middle of the turn, you could increase cross weight or decrease stagger. Also check for right rear tire pressure being too high.

Phase Two Handling Problems
Overster

Now let's look at phase two — the middle of the turn. If the car is loose here, look at these adjustments available to you:

(1) Increase the cross weight.

(2) The right front spring rate is too soft.

(3) The right rear spring rate is too stiff.

(4) The front sway bar needs to be stiffer or have more preload.

(5) The right rear tire pressure is too high.

(6) Decrease stagger.

(7) Lower the rear roll center.

With corner exit oversteer, the primary problem is usually too much stagger. Also, cross weight may have to be added.

Many times a push at mid-corner or turn exit is caused by the driver's style or corner entry line.

Phase Three Handling Problems
Oversteer
(1) Stagger would be the primary thing to look at here. If the car is loose at turn exit, decrease the rear stagger.

(2) Increase the cross weight – but stay within recommended guidelines.

(3) Soften the right rear spring rate.

(4) Check if right rear tire is over-inflated.

Understeering Problems

For an understeering condition, do just the opposite of what we have discussed for oversteer. For example, at turn entry (phase one), understeer (or pushing) can be caused by:

(1) A right front spring that is too stiff.

(2) A right rear spring that is too soft.

(3) A front sway bar that is too stiff.

(4) Too much front sway bar preload.

(5) Too much cross weight in the chassis.

(6) There is roll understeer in the chassis.

(7) Too much right front tire pressure.

(8) Not enough stagger.

(9) Too much front brake bias.

(10) The rear roll center is too low.

Also, refer back to the end of the *Phase One* subchapter for information on right front camber and shock absorber stiffness.

It is also important to analyze if the driver is causing the push by his driving style. For instance, braking hard into a turn with your left foot on the brake while you still have your right foot slightly on the gas will cause the car to push. This prevents the car from wanting to turn because the front end is settled down while the throttle is still pushing the car forward.

Basic Chassis Adjustments

Assuming that the race car comes to the track with the chassis set up properly (front end alignment correct, spring rates correct, weight distribution correct), then just five basic adjustments can be used to tune the chassis:

(1) Tire pressure

(2) Stagger

(3) Cross weight

(4) Rear roll center height

(5) Front-to-rear brake proportioning

Chassis Tuning With Air Pressure

Adjusting tire pressures is a good way to fine-tune the chassis. Tire pressures can change the car from push to loose, or vice versa. Air pressure is critical. A couple of pounds of pressure can make a big difference.

Adjusting The Stagger

The ultimate guideline for rear stagger is to use as much as is necessary to get the car to roll easily around the middle of the corner and not get loose on corner exit.

However, if you resort to large amounts of stagger to get the car through the turns, you are going to sacrifice turn exit traction plus increase tire wear. If your car requires too much stagger, try getting the car better balanced with different spring rates. If the

Measure tire pressure and circumference before the tires are used, then check them again after a number of laps have been run to determine tire growth and air pressure build up.

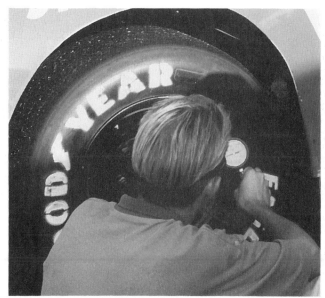

Be sure to constantly monitor stagger and air pressure when the car comes off the track. Keep good records.

car wants to push off the turns, reduce the right front spring rate, or increase the right rear spring rate.

Air pressure is going to affect tire stagger more than anything. For the feature race setup, you must know what the increasing tire pressure build-up is going to do to your setup. If you know your car handles best at 2.5 inches of stagger, but it increases to 4 inches through a feature race, the car will get so loose all you can do is hang on. You won't be able to race anybody. On the other hand, if the car loses stagger, like going from 2.5 to 1.5 inches, the car is going to push so much you won't be competitive. Be sure to measure tire growth and air pressure build-up over a set number of laps so you know where to start so that you end up with the ideal setup at the closing stages of a race.

Where you want your setup at the end of a race depends on the track surface history and the length of a race. Generally, you'll want to be tighter (pushier) for a longer race and a track surface known to get slick or slippery.

The key to going fast at any race track is to keep the momentum up through the middle of the corner. You are running on two different radii through a turn, and stagger is what will make a car run on the different radii. This will keep the car freed up so it will roll around a corner instead of having to be "driven" around the corner. This will allow you to

pick the throttle up without getting the rear end loose or creating a push.

Sometimes the amount of stagger used ends up being a compromise, depending on the length of the straightaways. For example, with long straightaways, too much stagger is going to make the car move around on the straights and get a little loose on corner entry, even though the car needs the stagger for a tight turn. The answer is a compromise between what is the fastest and what makes the driver the most comfortable.

Be sure to constantly monitor stagger and air pressure when the car comes off the track. Heat causes the inflation pressure to increase in the tires, and thus the stagger increases (with bias ply tires). Make a chart of tire pressures and tire circumferences each time the car comes off the track. This will help you determine how to set pressures for maintaining your optimum stagger.

Stagger versus crossweight is also a very important consideration. A car with a high amount of cross weight is affected more by stagger than a car with less crossweight. A car with a very heavily loaded left rear wheel is more critical for stagger. If it is off some, it will really be detrimental to handling.

Basic Guidelines For Using Stagger

1) More stagger makes a car looser coming off a turn and heading into a turn.

The caliper type of stagger gauge is the most accurate way of measuring stagger.

2) Less stagger tightens up the rear of the car.

3) If a car pushes coming off a corner, add more stagger.

4) The less rear weight percentage a car has, the less stagger the car needs.

5) The tighter the turns (shorter turn radius), the more stagger required. For example, the minimum stagger required with a 64-inch rear track on a tight 1/4-mile track might be 3 inches, but the same car on a wide sweeping 3/4-mile track (with the same banking) might be only 1.2 inches.

6) The flatter the track, the more stagger required. The more banking a track has, the less stagger needed.

7) If the track has multi-groove turns, the driver has to decide if he is more competitive driving the lower or the higher groove. This makes a difference in the amount of stagger to be used. Using the lower groove means the car will be using a tighter turning radius, which requires more stagger. The higher groove will not require as tight a turning radius, and thus less stagger is required. If the rear end is equipped with a locking spool, this stagger difference is critical. If a Detroit Locker® or some type of torque sensing differential is used, the car setup can be more tolerant of the differences in stagger for the different grooves.

8) More crossweight requires more stagger to get the car into and out of the turns. Conversely, less cross weight requires less stagger.

The flatter the track, the more stagger required.

If a track has multi-groove turns, the driver has to decide if he is more competitive driving the lower or the higher groove. There is a difference in turn radius between the two, which makes a difference in the amount of stagger used.

9) A race car is very sensitive to small changes in stagger. If your car does not react to small stagger changes, there are some other basic problems in the chassis. Don't try to correct them with massive amounts of stagger. You will only compound your problems. Make sure the basic points in the chassis are in order first – spring rates, alignment, weight distribution, etc. Then, and only then, do you use stagger to fine tune the chassis.

10) When picking out unmounted tires, keep in mind that the tires will grow about .75 to 1.0-inch after they are mounted.

11) To get a good indication of how a tire will grow in circumference when run, mount the tire and increase the right side tires to 50 PSI and hold it there for 20 minutes (first be sure that it is safe to increase

the tire pressure that high). Then bleed the tires down to 30 PSI and immediately take a circumference measurement. That number should give you a good indication of the final tire dimension after they are run. Mark the sidewalls with the circumference measurement. Different bias ply racing tires will grow at different rates, even though they are theoretically produced equally. (Radial tires are a completely different story. Because of their type of construction, they produce very little circumference difference between different tires.)

12) For an indication of tire growth of left side tires, follow the above procedures, but inflate the tires to only 30 PSI, hold for 20 minutes, then deflate them to 20 PSI and measure the circumference.

13) Always keep records on stagger. Measure the tires' circumference before and after each time the car races an event – heat races and main events. Then if a handling problem develops, the first thing you should look at is the tire stagger. Has it changed? In many cases, a change in the air pressure and stagger is what has caused the problem.

Cross Weight

More cross weight is used to keep the rear end of the car tight on corner entry, and to improve bite off the corners. It favors the left rear tire contact patch by more heavily loading it.

The total amount of cross weight required depends on track size and configuration, and the amount of roll couple used in the car.

A car with a high horsepower/higher torque engine will usually use more cross weight, because the high power creates power oversteer. A lower horsepower car will require less cross weight because it doesn't have the power to break the rear tires loose and cause oversteer.

Stagger Vs. Cross Weight

Stagger and cross weight are two very separate and distinct chassis adjustment tools, yet when one is changed, the other is affected also. Be sure to keep this in mind when changing stagger. Going to a smaller circumference tire at the right rear (less stagger) will increase the cross weight, or going to a larger circumference tire at the right rear (more stagger) will decrease cross weight. This is because a larger right rear tire raises that corner, taking weight

With the chassis mount end of the Panhard bar attached to an adjustable bracket or clamp-on bracket, the rear roll center height can easily be moved up or down.

off of the left rear and adding it to the right rear. Conversely, using a smaller right rear tire lowers that corner of the car, which adds weight to the left rear.

Changing stagger is a more chassis-sensitive adjustment than changing cross weight. In other words, stagger can make a more precise, sensitive change in the car's handling. This is because stagger changes the steering effect of the rear end (stagger makes the right rear want to steer about the left rear) whereas cross weight increases the force downward on a tire (no steering effect). A one-inch difference in tire circumference makes about a 3/16-inch difference in tire height. So, if you install a tire on the right rear which is two inches in circumference greater, you have raised that corner of the car 3/8-inch (2 times 3/16-inch). That means you have taken 3/8-inch of wedge out of the car. As you can see, the change in stagger has a much more dramatic effect on the car's handling than does the cross weight change. When dramatic changes in stagger are made to the chassis, cross weight will have to be restored to keep the chassis balanced.

Chassis Tuning Using The Rear Roll Center

The rear roll center is that point about which the sprung weight of the chassis transfers from inside to outside during cornering. It is controlled by the Panhard bar on coil spring and coil-over cars.

The length of the lever arm between the rear roll center and the center of gravity height will influence the amount of body roll taking place at the rear. By raising or lowering the rear roll center, the amount of body roll is changed. A longer lever arm (lower roll center) creates more body roll. The more body roll present, the more the transferred weight during cornering will be placed on the outside tire as vertical loading. The vertical loading onto the tire contact patch will create side bite and forward bite.

In situations where the Panhard bar can be raised or lowered, this adjustment distributes between vertical loading and horizontal loading of the right rear tire contact patch during body roll. A lower rear roll center will load the transferred weight more vertically onto the tire and tighten up the chassis. A higher rear roll center will load the tire more horizontally and loosen up the chassis during cornering. With the chassis mount end of the Panhard bar attached to an adjustable bracket or clamp-on bracket, the rear roll center can easily be moved up or down.

Effects Of Spring Rate Changes

Changing spring rates at the right front and right rear will change the roll couple distribution of the chassis. For example, changing to a stiffer right front spring will make the car tend more toward understeer, or changing to a stiffer right rear spring will make the car tend more toward oversteer. But, changing spring rates on the left side of the chassis will also have an influence on the car's handling characteristics.

A softer spring rate used at the left front will help to tighten up a chassis at turn entry and into the middle of the turn. A stiffer left front spring makes a car looser going in and coming out of a turn because it subtracts cross weight from the chassis.

A stiffer left rear spring rate tightens up the chassis from the middle of a turn and through the exit by keeping the cross weight in the chassis.

Lower Trailing Arm Angles

Raising and lowering the front of the rear trailing arms has a direct effect on the amount of load placed on the rear tires during acceleration. A trailing arm uphill angle adds more tire loading under acceleration. This is because the rear end is trying to move up underneath the chassis as it pushes these links

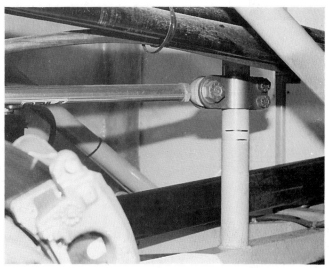

Clamp brackets are an effective way of making small angle changes with lower trailing arms. Note how this chassis has been marked for the correct setting.

forward. The upward-angled arms are reacting against the weight of the chassis which provides more tire loading. This is called axle thrust.

Raising the front of the trailing arm increases axle thrust and tire loading. So, for example, if you need more left rear tire traction while accelerating off the turns, raise the front of the left rear trailing arm. But, you have to remember that the loading produces an equal and opposite effect under deceleration at turn entry. It will tend to take load off of the left rear tire and loosen up the car.

Uphill angle of the lower trailing arms can also cause roll steer problems during cornering. The best answer is a compromise. A slight amount of uphill angle can be used, usually between 1 and 3 degrees. A very common setting is 3 degrees uphill at the left rear, and 2.5 degrees uphill at the right rear. Be sure the car is setting on your level setup pad when making these adjustments.

Rear End Tracking

Rear end tracking is the lateral positioning of the rear end housing as compared to the front wheels positioning. The rear end housing can be pulled to the left or to the right, in relationship to the chassis, by lengthening or shortening the Panhard bar.

The tracking is used to gain or reduce traction while accelerating off the turns. If the rear end is moved over to the right, it makes the car looser as it accelerates off the turns. If it is moved over to the

REAR END TRACKING

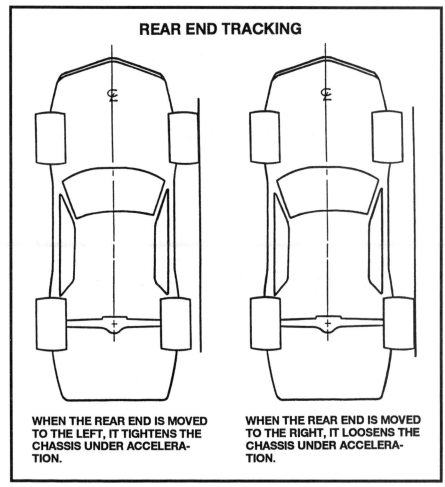

WHEN THE REAR END IS MOVED TO THE LEFT, IT TIGHTENS THE CHASSIS UNDER ACCELERATION.

WHEN THE REAR END IS MOVED TO THE RIGHT, IT LOOSENS THE CHASSIS UNDER ACCELERATION.

The banking and downforce will also increase rear tire bite. To overcome this, the rear spring rates have to be increased, especially at the right rear. If you have a problem getting the car to turn, increase the right rear rate more. You can fine-tune the chassis by adding more stagger (but don't use stagger as a crutch to cure other problems, such as spring rates).

The anti-roll bar diminishes its rate effectiveness the more that the track is banked. This is because of the downforce loading on the chassis caused by the banking. Banking will cause more downward loading on the chassis and less body roll, and the anti-roll bar is only effective during body roll.

Be careful of getting the overall chassis setup on a banked track car too soft. A soft setup has more body roll, and banked tracks are faster and make things happen quicker. The result is that a soft setup can end up reacting too quickly.

left, it tightens the car up as it accelerates, because the offset housing is trying to steer the car to the right.

This same effect can be achieved with the use of wheel spacers.

Left offset tracking is used in a situation where the track is very slick and it is hard to get good traction as the car accelerates off the turns. Right offset tracking can be used when the car is very tight coming off the turns due to very sticky tires or very little horsepower, and there is no other way to loosen the car up.

Rear end tracking is generally a tuning technique that is used when the other chassis adjustments have not produced the desired handling effects.

Banked Track Corrections

Going to a medium-to-higher banked track from a flat-to-slightly banked track requires several changes in the chassis set-up. First, banking creates downforce, so all four spring rates have to be stiffer to resist it and hold the car up.

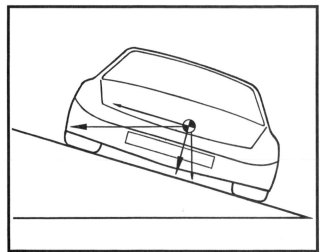

On a banked track, force vectors turn sideways centrifugal force into downward centrifugal force. This downward loading pushes the chassis down equally on both sides and thus diminishes body roll. For this reason, an anti-roll bar's rate is much less effective on a banked surface.

Using an anti-roll bar to adjust roll couple is one of the easiest and most effective adjustments to sort out oversteer or understeer.

For a higher banked track, you want a lower front roll center and less negative camber gain, because there is much less body roll and there is more downforce. On a flatter track you have much more body roll, so you need more negative camber gain which you get with a slightly higher front roll center.

Paved Track Chassis Adjustments

The first thing you have to do is decide the cause of a handling problem. For example, let's say you are racing on a 3/8-mile flat paved track, and your car is equipped with a spool in the rear end. There is very little rear tire stagger. The car experiences a push in the middle of the turn. The source of the push is probably the spool in the rear end which keeps the rear wheels solidly locked together, driving the front end straight ahead rather than allowing it to be steered. In this case, the rear end needs some help getting the car steered in the turn. The answer is to add some tire stagger to help steer the rear in the turn.

Now, let's take the same car, track and situation again. But the car already has 4 1/2 inches of stagger and it still pushes at turn entry and in the middle of the turn. We have to remember that the car is already at — or beyond — the suggested maximum amount of stagger, so the problem must be with roll couple distribution. That means the front end roll resistance is too stiff. If the push is slight, the best change would be to take some preload out of the front sway bar. Or, you could take some cross weight out of the chassis. If it is a little more severe push, change the

front anti-roll bar to the next lightest size, or adjust the bar arm length so it is longer (which makes the bar effectively weaker). Using the anti-roll bar to adjust roll couple is one of the easiest and most effective adjustments to sort out oversteer or understeer. It doesn't change any suspension heights or the geometric relationship of one suspension piece to another.

If the push condition is fairly severe, then a right front spring change would be in order (go to a lighter spring rate). Or, the right rear spring can be changed to a stiffer rate.

What if we have a fairly severe push, and the decision is made to make a spring rate change? Do we soften the right front spring, or stiffen the right rear spring? Either one would balance the car's handling. However, choosing the incorrect answer will decrease the maximum achievable lateral acceleration. We have to look at tire temperatures to make the correct choice. If the right front tire temperatures are higher than normal, or much higher than the right rear temperatures, then this information points to the problem being at the right front. We would want to decrease the roll resistance at that corner, or in other words, install a softer rate spring there. If, on the other hand, the tire temperatures show the right front temperatures within the optimum operating range, but the right rear temperatures cooler than normal, then the problem of understeering is caused by too little load being handled at the rear of the car. In this case, the right rear spring rate would be stiffened. This would increase the roll resistance being handled by that tire, and bring its operating temperature up.

If the problem is one of severe oversteer, the same procedure of checking the operating range of the tires would be followed. Balancing the chassis by bringing the operating temperatures of the front and rear tires within optimum range will maximize the lateral acceleration capability of the race car. When the chassis is perfectly balanced, on a flat to low-banked 1/4-mile to 1/3-mile paved track, the right front tire temperature average will be about 10 to 15 degrees hotter than the right rear tire temperature average. Any more heat than this at the right front indicates pushing (or understeer). Any less heat at the right front indicates oversteer or looseness.

When sorting the chassis for a particular problem, we must isolate where on the track that problem

If a car is loose under power from the middle of a turn and through corner exit, the problem might be able to be cured by adding more sway bar preload if the problem is slight. If it is more severe, work with tire stagger.

If a car is loose at corner entry, the driver might use an early apex, which would make the car push at mid-turn and corner exit. The car has to be adjusted to handle properly at corner entry.

occurs. For example, we have a push. Where does that push occur — at turn entry, in the middle of the turn, or through the exit? A push in each position is a symptom of a very different problem. Let's look at each one:

Push at turn entry When this happens during braking, the problem is many times a brake proportioning problem — too much front brake bias. Adjust the proportioning so that the rear brakes are doing a little more of the work.

The opposite is also true. If the car gets loose at turn entry under braking, the first thing to suspect is too much rear braking bias. If you suspect this, try entering the turns without using the brakes. This will tell you if oversteer or understeer is present in the chassis without the influence of the brakes.

Push in the middle of the turn This is most probably a tire stagger problem. Increase the stagger just a little.

Push under acceleration off the turn There is probably too much cross weight in the chassis, or too much left rear weight. However, if the car starts to push in the middle of the turn and consistently continues all the way onto the straightaway, the problem is probably not enough stagger. Increase the stagger just a little.

All of these chassis adjustments discussed above assume that all else in the chassis — springs, shocks, anti-roll bar, weight distribution, etc. — are correct

already, and we are talking about fine tuning changes.

If the car is loose instead of pushing, we again have to make the determination where the problem occurs on the race track in order to make the proper chassis adjustment. Let's look at each case:

Loose during entry into turn This is probably caused by too much stagger or too much rear brake bias.

Loose under power from the middle all the way off the turn To diagnose this, you need to determine how excessive the oversteer is. With just a little looseness, the problem could be cured with a sway bar adjustment. Use more sway bar preload to add more cross weight in the chassis. With a lot of oversteer, the problem is most probably too much tire stagger. Go to a smaller circumference right rear tire, or a larger circumference left rear tire.

Again, in all of these discussions, we are assuming the problems here are fairly minor, requiring only fine tuning, and that the chassis is already in the ballpark with springs, shocks, alignment, weight distribution, etc.

Loose going into a turn, pushing under acceleration coming off the turn Like all handling problems, always start by solving the problem which occurs at turn entry first. Begin by ruling out a problem of too much rear brake bias making the car loose at turn entry.

Use a slightly stiffer right front spring rate to tighten up the chassis. Alternatively, a slightly softer left front

When trying to diagnose handling problems, be sure that you have good records of tire pressures and tire temperatures. This will give you a good indication of what the chassis is doing and where the problems might be.

spring would help to tighten up the chassis at turn entry.

The stiffer right front spring will not necessarily make a chassis tighter all the way around the race track. The situation might really be that the car is loose going in and coming out, and the driver is driving it in such a way that it will tighten up in the middle and coming off. He could do that by taking an early apex at the corner entry. With the car consistently loose, the driver does not have the confidence to take the proper line and turns the steering wheel a little more to the left at turn entry.

Tightening the chassis up at turn entry will, in a lot of cases, fix the problem at turn exit because the driver is able to drive the car on the proper line into the corner and through the middle.

If the car is still tight at corner exit, there are several things to do to loosen it up at this point. You could take out a slight amount of cross weight – such as half a turn at the left rear. You could increase tire stagger slightly by adding a pound or two in the right rear tire, or by taking a pound or two out of the left rear tire.

While trying to diagnose this handling problem, be sure that you have good records of tire pressures and tire temperatures. This will give you a good indication of what the chassis is doing. If you get a lot of pressure build-up in one tire, that tells you the car is loose or pushing. More build-up at the right rear is

caused by a loose condition. More build-up at the right front is caused by pushing. Tire temperatures will confirm this.

Remember, the suggested chassis adjustments discussed in this segment are fine-tuning adjustments only. Before they work effectively, the proper spring rates and weight balance must be in the car. If the car's handling is really off, work to get these items right first.

Trouble Shooting Guide

Car Is Loose Going Into A Turn

Item	Adjustment
Stagger	Right rear is too large
Toe-out	Not enough
Front springs	Too weak
Rear Springs	Too stiff
Brake bias	Too much rear
Suspension bind	Exists in rear suspension
Cross weight	Increase

Car Pushes Off Corner

Item	Adjustment
Stagger	Increase right rear size
Toe-out	Decrease
Front springs	Too stiff
Rear springs	Too weak
Cross weight	Use less

Note: This is a very basic quick reference guide to common handling problems. The charts are not meant to cover all problems and situations, but rather to give you an idea of where to start when you experience a problem.

Shock Absorber Troubleshooting

Many times all indications would show that the chassis is set up correctly, and yet the driver experiences "transient" or momentary handling difficulties with his car, particularly at just one point on the track, such as going into a turn, or during acceleration from corner exit. A problem of this nature points to an improper shock absorber rate at one or more corners of the car. The following is a discussion of the particular problem caused by a shock which is too stiff or too soft at one corner of the race car. Each corner diagnosis here assumes that the shock causing the problem is the only improper rate shock on the chassis — all others on the car are correct for the application.

Shock Absorbers Too Stiff

Right Front A right front shock which is too stiff slows down the rate of body roll and weight transfer. It causes turn entry pushing because the right front corner of the car is momentarily over supported with too stiff a shock, and the tire is overloaded.

Left Rear If the left rear shock is too stiff, it keeps the tire hooked to the chassis as it unloads during turn entry, because the shock is very slow to extend, thus the wheel is pulled up temporarily. By losing the left rear tire contact patch, the right rear is overloaded during initial turn entry and the rear end gets loose.

Right Rear If the right rear shock is too stiff, there is too much initial body roll resistance, and it causes the right rear tire to be overloaded momentarily.

Left Front If the left front is too stiff, on turn entry the shock pulls that tire up with the chassis during body roll, partially unloading that corner. This causes an overload of the right front, and thus a momentary understeer.

Shock Absorbers Too Soft

Right Front If the right front rate is too soft, the shock can't handle the cycling of the spring effectively, and that corner of the car bounds up and down. Too soft of a right front shock also allows weight to be transferred very quickly to that corner on turn entry, causing a momentary push.

Left Front If the left front shock is too soft, it loads the right front too quickly and causes an initial push on turn entry.

Right Rear Under-control of the spring action at this corner causes a bounding of the car at this corner. It causes an instability of the tire contact patch, and makes the car transition quickly from push to loose and back again. Too soft a shock rate at the right rear can allow the car to pick up the left front wheel during acceleration at the corner exit.

Left Rear Under deceleration, too soft a shock rate at the left rear causes too quick a pitching moment to the right front, which overloads that tire and causes a push.

Under deceleration, too soft a shock rate at the left rear causes too quick a pitching moment to the right front, which overloads that tire.

Also, when we talk about the symptoms of a shock absorber being too soft for a particular application, don't overlook the fact that the shock rating might be correct for the application, but the symptom is being caused by a bad shock.

Do not hesitate to replace a shock absorber that you suspect is damaged. Improperly functioning shocks can cause all kinds of unknown handling maladies. For example, a slightly bent shaft on the right front shock will cause a push on turn entry, or a bent shaft at the right rear would cause on an oversteer (because it would keep the shock from compressing properly).

Another problem that happens periodically in double tube shocks is that the gas bag breaks (although they are really quite durable). When this occurs, the shock fluid mixes with the nitrogen gas and aeration results. The shock fluid is then compressible and the shock offers little resistance to movement.

Shock Travel Indicators

Shock travel indicators can be a good "rule of thumb" check for a chassis to see if it is handling as it should, and if not, which end of the car to look at to find the problem. When the travel at one corner is out of line in comparison to the others, it points immediately to the corner of the car where the problem lies.

Shock travel indicators are a good ballpark reference point for any type of race car, regardless of weight or weight distribution because the chassis

One of the most important aspects to understand about shock absorbers is that rebound control is 75 percent of what makes a car faster. Stiffer rebound control on the left side of the car slows weight transfer from inside to outside during turn entry, allowing a car to drive deeper into a turn.

will be adjusted accordingly to compensate for differences (such as a heavier car will have proportionally stiffer springs).

Good ballpark shock travel numbers to look for would be:

Right Front:	**1 1/2 to 1 3/4 inches**
Right Rear:	**2 to 2 1/4 inches**

If you want to use shock travel indicators to give you some meaningful feedback on your chassis, be careful that the car does not run over any bumps or potholes that would cause abnormal shock movement, thus giving you false indicators.

The amount of shock travel your car experiences, in relation to the ballpark numbers quoted above, is going to depend a great deal on where the shocks are mounted on your suspension (how close to the wheel) and at what angle the shock is mounted. The greater the distance away from the wheel the shock is mounted, and the greater the angle away from vertical, the less total indicator travel the shock will experience.

Wheel travel is not the same thing as shock indicator travel. This is because the shock absorber is mounted to the inside of the wheel, and also at an angle. You could mathematically compute wheel travel from shock travel, but there's an easier way. Block the right front of the car's frame at the proper ride height (to check right front wheel travel), then remove the spring. Jack up the right front suspension until the shock absorber grommet moves the same amount as it normally does on the race track. Measure the distance you jacked the wheel, and you now know how much wheel travel your car has.

How much wheel travel should your chassis experience? On a paved track, approximately 3 inches at the right front, and 2 inches at the right rear.

Chassis Tuning With Shock Absorbers

Racers who understand shocks and how to use them to fine tune their chassis will have a distinct advantage. Racers are always looking for an edge, and they can find it through tuning with shock absorbers.

The rate of shock absorber damping influences dynamic weight transfer. Shock absorber valve damping characteristics determine how quickly a car unloads weight from left to right, or front to rear, or vice versa. They also control diagonal pitching moments from the left rear to the right front under braking and the right front to left rear pitching under acceleration. Control of dynamic weight transfer during cornering is determined by low and mid range shock valving.

One of the most important aspects to understand about shock absorbers is that rebound control is 75 percent of what makes a car faster. Stiffer rebound control on the left side of the car slows the weight transfer from inside to outside during turn entry. This allows a car to drive deeper into a turn and stay balanced because there is more weight for better bite remaining on the left side.

Split Valving Shocks

A split valve shock has a damping rate in rebound control that is different than its damping rate for compression. Split valving shocks can be used to fine tune specific corners of the race car for specific handling situations.

A shock that is stiff on rebound valving and softer on compression valving is called a "tiedown" shock. It keeps the corner of the car where it is attached tied

If a vehicle is pushing at corner entry and using a 95 shock at the left rear, change to a 956 at the left rear. The stiffer rebound slows transitional weight tranfer to the right front at corner entry.

down, making it very hard for the body to rise up at that corner.

There are varying degrees of tie-down shocks. For example, a 938 is a 9-inch stroke shock which has a 3 valving in compression and an 8 valving in rebound. (The shock numbers we quote here are always in that sequence – stroke length, compression valve code, rebound valve code All manufacturers except Carrera list them in that order.) This is a severe tie-down shock — it is very soft in compression and very stiff in rebound. A 956 is also a tie-down shock, but it is not nearly as severe in its split valving. It has a 5 valving in compression and a 6 valving in rebound. You have to know how to choose between using these, depending on track conditions. We'll explain that shortly.

The opposite of a tie-down shock is an "easy up" shock. It has a rebound rating that is soft, and a compression that is a stiffer valving. The easy-up shock is used on a corner of the car when you want to have weight transfer very quickly from that corner to the opposite end of the car. For example, on a slick race track, on acceleration you want to have weight transfer from the front to the rear wheels very quickly. The easy-up (soft rebound) shock is used here.

How these shocks can be used to remedy transitional handling problems is shown in the following examples:

Loose At Corner Entry

If the looseness is not a brake proportioning problem, and the car transitions to neutral handling at mid corner, then a shock absorber can help the chassis here.

If a 95 shock is being used at the left rear and right rear, change to a 954 shock at the left rear. This is a split valve shock that is one step softer in rebound. This helps the weight to transfer from the left rear to the right front tire quicker. It also has an influence on the right rear tire, allowing the weight to roll over to the right rear quicker, which gives it more bite (provided the rear roll center height is correct).

Pushing At Corner Entry

Shocks can help here if the problem is slight and the push occurs during the transition from straightaway to cornering.

If the car is using a 95 shock at the left rear, change to a 956. This is a split valve shock that is one step stiffer in rebound. What this does is slow the transitional weight transfer from the left rear corner to the right front during corner entry. It prevents weight from transferring too fast, which would overload the right front tire contact patch. Using the split valve shock at the left rear, the 5 valving is still retained in compression, which is what the chassis requires for good hook-up under acceleration at corner exit.

Loose At Mid Turn

There are several items which could cause the car to get loose at mid turn It could be spring rates, or it could be that the tires are not getting enough grip on the right side. This happens when there is not a long enough lever arm between the roll center and the CGH. In this case, the weight is transferred more as shear force at the tire contact patch rather than down force, which generates grip. To remedy this problem, the lever arm distance has to be increased by lowering the Panhard bar, which lowers the rear roll center.

If the problem is just a slight transitional looseness at mid turn, the problem can be handled with a split valve shock at the left front or right rear. A softer rebound valving at the left front allows the car to roll over quicker onto the right rear, which sticks the right rear tire and gives that tire more down load for better grip. If you are using a 76 shock at the left front, switch to a 765.

If the car is pushing at mid-turn and a 76 shock is being used at the left front, change to a 767. The increased rebound damping will keep the weight on the left front longer.

Preload is added to the anti-roll bar by raising up on the bar's load bolt on the left side or by lowering the load bolt on the right side.

Pushing At Mid Turn

This problem occurs when the left front corner is not doing enough work to help hold the car on its line. With more weight on that tire, the left front will improve directional control. If you are using a 76 shock at the left front, change to a 767. The increased rebound damping will keep the weight on the left front longer. It prevents the car from rolling onto the right side – especially the right rear — too quickly, which helps the left front work better.

Loose At Corner Exit

If the car is neutral at corner entry and through the middle, but it gets loose as the driver starts to accelerate off the corner, the chassis needs to transfer weight quicker onto the left rear corner. This is accomplished by softening the rebound damping at the right front. Changing to a 764 shock at that corner allows quicker transitional weight transfer under acceleration to the left rear, putting more downward weight loading on the left rear tire.

Push At Corner Exit

Change to a stiffer rebound valving at the right front, or a stiffer compression valving at the left rear. This slows the transitional weight transfer to the left rear under acceleration, which allows the right rear to have slightly more driving force. This keeps the car loosened up at turn exit at the point where the driver begins his acceleration.

The Driver's Effect On Shock Choice

A racer's driving style can have an influence over which shocks are used at each corner of the car, or which ratio of compression/rebound damping is used. Abrupt versus smooth application techniques of the steering, gas and brake make the difference between faster and slower shock absorber shaft speed, which can mean different damping characteristics are required for different drivers.

Chassis Tuning With Front Anti-Roll Bar

Adjuster bolts that are part of the front anti-roll bar mount can be used to add preload into the chassis. Adding preload to the front anti-roll bar tightens up the chassis while the car is entering and exiting the turns. If the bar rate (instead of preload) is changed, it will affect the handling of the car all the way through the turns – not just at corner entry and exit.

Usually a chassis is set up with the anti-roll bar neutral (no preload), or up to 1 percent preload. This is set when the car is still on the wheel scales. The preload adjuster is screwed down until the scales show that 1 percent more cross weight has been added.

Preload is added to the bar by raising up on the bar's load bolt on the left side or by lowering the load bolt on the right side.

Be sure to make a record of how much the cross weight is changed as you make changes to the load bolts with the car setting on the wheel scales. If you know how many turns it takes to increase the cross weight 1 percent, it helps you to make a quick positive adjustment at the race track without having to guess what you are doing.

Weather And The Race Track

On a paved track, sun and heat make a race track slicker. The warmer the track and air temperature, the less traction that is available. This happens because oil, grease and moisture are driven to the surface of the pavement by warmer temperatures. The warmer the temperature, the slipperier the track, and the less traction that is available. Different tracks react differently to heat, but in general the loss of traction caused by heat will require the chassis to be tightened up. Tighter means that the car will tend more toward understeer. Generally, the cross weight can be increased to adjust for this change in conditions.

A cool paved track surface contracts and makes it slightly more abrasive. This allows the tires to grip better. When the track surface gets cooler, a slightly looser chassis set-up is required.

A different type of condition occurs on asphalt track surfaces early in the season after the track has been subjected to harsh Winter weather. Over the Winter the pavement has been subjected to rain, snow and freezing temperatures. As the track opens for racing in early Spring, the pavement is slick and polished from the weather. There is no rubber laid down. It is difficult to get a good grip on the slick surface. The chassis setup for this track condition is going to be entirely different than in the heat of Summer.

This type of surface condition will require a softer tire compound (if no spec tire rule is in effect), softer spring rates, a slightly higher ground clearance (to get more weight transfer onto the outside tires), and a stiffer front anti-roll bar rate. The softer front springs will help the car to turn into the corners, and the

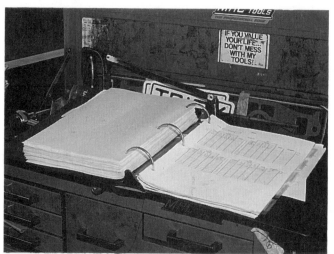

A good set of records will help you tremendously in deciding what to do when diagnosing handling problems. This loose-leaf book is one competitor's set of records for his car for just one season of competition.

stiffer front bar will keep the car from falling over onto the right front. Under acceleration, the bar helps to accelerate weight transfer back to the left rear as it unwinds to keep the chassis tight.

For this type of slippery surface condition, the rear shocks should be one step softer in compression. The softer bump damping accelerates weight transfer onto the rear wheels under acceleration, which helps prevent wheel spin on the slick surface.

The important thing is to closely watch your own particular track week after week. Keep notes so that you can remember and compare. You need to make notes about what chassis changes were required if the weather was very hot, and how the car reacted to these changes. Keep a record of ambient air temperatures and track surface temperatures. The amount of humidity in the air may also influence how the car handles.

What you want to do is know the tendencies of your track so you know what to expect. Knowing what to expect can give you a tremendous advantage in setting up your car for the main event.

Roll Couple Distribution

Roll couple is the force, due to cornering, acting on the sprung weight of the car, rolling it about the roll axis. It is the action that transfers weight from the inside wheels to the outside wheels during cornering.

The front-to-rear handling characteristics (understeer or oversteer) of a race car can be tailored or adjusted by adjusting the front-to-rear roll stiffness proportioning. This is called roll couple distribution. The end of the car with the highest roll stiffness will receive the greatest amount of weight transfer during body roll. The higher the front roll couple percentage, the stiffer the front roll stiffness is, which means more of a tendency toward understeer. Conversely, the less front roll couple, the more oversteer.

Making A Spring Rate Change At The Track

Make a spring rate change only after all other adjustments have been exhausted. That is because making a spring rate change at the track can many times cause more problems than it solves if the proper procedure is not carefully followed.

Before the spring is changed, measure the compressed height of the spring you are about to change so that the chassis can be adjusted back to this same height. If you are using a coil-over unit, measure the installed center-to-center length between the upper and lower coil-over mounting bolts before the spring is changed. The compressed height of a spring will change because a softer spring will compress more, or a stiffer spring will compress less. If the change in height is not adjusted, you will end up with weight jacked into or out of the chassis.

Make the spring rate change, and then adjust back to the correct compressed height. Then double check your frame heights at all four corners of the car to be sure the new spring and your readjustment for compressed height of the spring has not changed anything else. You may have jacked some weight into or out of the chassis without realizing it. This is the part that makes a spring change at the track the toughest. There is seldom an adequate flat surface to measure to in the pits. Your best procedure — if the pit surface is not a flat paved surface — would be to put a flat board on the ground underneath each position where you measure frame height, and take a measurement at each corner from the frame to the board before you make the spring change. Then, with these boards still in place, you have a flat reference surface to make measurements to.

After you set up the chassis at the shop, you should have measured the length of each shock absorber (mounting bolt to mounting bolt centers) with the car setting at the correct ride height, and recorded the numbers. Then use these numbers to relevel the chassis should a spring change be made.

Rented Track Or Skid Pad Testing

One of the most effective things you can do to get your chassis dialed-in is to rent a race track or a skid pad, and spend the day testing and making changes.

To get the most out of any test session, you have to show up with the car totally prepared. This means having the correct tires, stagger, gearing, ride height, alignment, and the weight distribution set at your best estimation of proper left weight, rear weight and cross weight. And by all means, have the complete setup specifications recorded. This eliminates any guess work when making changes, and saves time at the track. You will be wasting your time if you don't know the baseline you are starting with.

Being prepared also means bringing all the necessary tools and parts with you. Be sure to have every tool necessary to make any change that may be necessary. This should include measurement tools such as toe gauge, stagger gauge, and tire pyrometer, and an air tank or air compressor to change tire pressures. Bring all of the various front and rear springs you want to try, various front and rear shocks, anti-roll bars, and several different circumference tires so stagger can be tested.

Also, be sure to have various replacement parts along with you in case you suffer engine or driveline failures, etc. Think of it as going to a race – be ready for any situation. If you have engine problems during a test, if you don't have the tools or parts necessary to correct the situation, it will cost you a very important information gathering session.

The records you keep during the testing sessions give you a baseline to go back to, and knowledge to build on. Be meticulous in your record keeping. Record everything. You cannot save enough information. Even if you think it may be insignificant, record it any way.

Treat any testing session just like a race. The driver should wear full safety gear including firesuit, racing shoes and gloves, and helmet. He should also wear full safety belts and use the window net. Have fire extinguishers standing by in the pits and near the track. Just because it is a test rather than an actual race, there is not reason to skimp on safety. Remember that the race car will be operating at racing

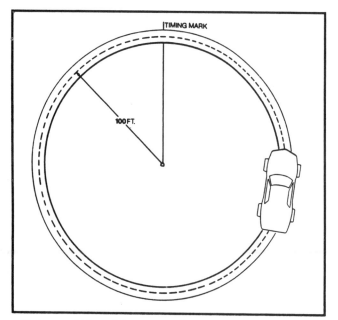

Using a skid pad is an excellent way of testing the handling of a car. The car is driven at a steady state as fast as possible, centered on the circumference of the skid pad circle.

speeds, and there is always a possibility of a crash, a part failure, a fire, etc. Do everything possible to protect the driver.

Testing Procedure

The car should arrive at the track or skid pad fully prepared. Make sure you have the complete baseline setup recorded. Record air and track surface temperatures. Check tire pressures and record them. Then run 10 laps to settle the chassis, and get heat in the tires.

Pull the car off the track very quickly (or stop on the straightaway by the flag stand) and quickly record tire temperatures.

The initial tire temperatures give you an indication of what the chassis is doing. Analyze the problem, and start making changes. Remember to only make one change at a time. If you make 2 or more changes, you will not be able to positively identify which change created which result. Then continue testing your changes in 10-lap segments.

Remember to record track surface and air temperatures every half hour so you know how temperature changes affect the handling or test results.

After each testing segment, record tire temperatures, tire pressures and tire circumference measurements.

Suppliers Directory

AFCO Racing Products
977 Hyrock Blvd, P.O. Box 548
Boonville, IN 47601
(812) 897-0900 Fax (812) 897-1757
www.afcoracing.com
Coil springs, shock absorbers, all types of suspension components

Appleton Rack & Pinion
110 Industrial Drive, Bldg. E
Minooka, IL 60447
(815)467-1175 Fax (815)467-1179
www.appletonrackandpinion.com
Power rack and pinion steering systems, suspension components

Carrera Shocks
QA1 Precision Products
21730 Hanover Ave.
Lakeville, MN 55044
(800) 721-7761
www.qa-one.com
Shock absorbers, coli-overs, springs

Coleman Machine
N-1597 US 41
Menominee, MI 49858
(906) 863-7883 Fax (906)863-7027
www.colemanracing.com
Aluminum spindles, suspension components

Crow Enterprizes
418 E. Commonwealth Ave., Suite 1
Fullerton, CA 92832
(714) 879-5970
www.crowenterprizes.com
Safety equipment

Dan Press Industries / Sierra Racing Products
4220 Petaluma Blvd. North
Petaluma, CA 94952
(707) 283-4DPI (4374) Fax: (707) 762-6961
www.dpiracingproducts.com
Disc brake systems, Gold Trak differential

Five Star Stock Car Bodies
PO Box 700
Twin Lakes, WI 53181
(262) 877-2171 Fax (262) 877-3702
www.fivestarbodies.com
Stock car bodies

Fuel Safe Fuel Cells
250 SE Timber Ave
Redmond, OR 97756
(541) 923-6005 Fax (541) 923-6015
www.fuelsafe.com
Fuel cells

Howe Racing Enterprises
3195 Lyle Rd.
Beaverton, MI 48612
(989) 435-7080 Fax (888) 484-3946
www.howeracing.com
Race car chassis, suspension components

JR Motorsports
801 SW Ordnance Rd.
Ankeny, IA 50021
(888) 771-5574
www.jrmotorsportsltd .com
Stock car parts

Kwik Change Products
P O Box 34260
Indianapolis, IN 46234
(317) 506-4838 Fax (317) 271-7908
Kwik Change tire pressure relief valves

Longacre Racing Products, Inc.
16892 146th St. SE
Monroe, WA 98272
(360) 453-2030 Fax (360) 453-2031
www.longacreracing.com
Gauge panels, switch panels, roll bar padding, chassis scales

Midwest Motorsports
2100 E. Lincoln Way
Ames, IA 50010
(800) 262-5033
www.midwestmotorsportsinc.com
Stock car parts

Motor State Distributing
8300 Lane Drive
Watervliet, MI 49098
800-772-2678, Fax 800-772-2618
www.motorstate.com
  Stock car parts

Outlaw Racing Disc Brakes
1465 Ventura Dr.
Cumming, GA 30040
(770) 844-1777
www.outlawdiscbrakes.com
  Disc brakes

Port City Racing
3011 Mill Iron Rd.
Muskegon, MI 49444
(800)4-RACING Fax (800)441-6875
www.portcityracing .com
  Race car chassis, suspension components

PRO Shocks
1715 Lakes Parkway
Lawrenceville, GA 30043
(770) 995-6300
www.proshocks .com
  Shock absorbers, coil-overs, springs

Quick Change Exchange
11775 Slauson Ave
Santa Fe Springs, CA 90670
(562) 696-4020
www.quickchangeexchange.com
  Quick change rear ends

Roehrig Engineering Inc.
633 McWay Dr.
High Point, NC 27263
(336) 431-1827 Fax (336) 431-7469
www.roehrigengineering.com
  Shock absorber dynos, shock dyno testing

Ron's Rear Ends
14640 Macneil St.
Mission Hills, CA 91345
(818) 361-5457
www.ronsrearends.com
  Ford 9" specialist

Speedway Engineering
3040 Bradley Ave.
Sylmar, CA 91342
(818) 362-5865 Fax (818) 362-5608
www.1speedway.com
  Quick change rear ends

Speedway Motors
300 Speedway Circle
Lincoln, NE 68502
(402) 323-3200 Fax (800) 736-3733
www.speedwaymotors.com
  They sell almost every part necessary for building a race car. Their catalog is a must

Swagelok Tube Fittings
Crawford Fitting Co.
29500 Solon Rd.
Solon, OH 44139
(440) 349-5934 Call for a distributor near you
www.swagelock.com
  Brake line fittings

Sweet Manufacturing, Inc.
3421 S. Burdick
Kalamazoo, MI 49001
(800) 441-8619 Fax (269) 384-2261
www.sweetmfg.biz
  Rack and pinion steering, power steering, spindles, steering u-joints, suspension brackets

Tex Racing Enterprises Inc.
2268 US Hwy 220 Alt. North
Ether, NC 27247
(910) 428-9522 Fax (910) 428-1734
www.texracing.com
  Rear ends, driveshafts, race-prepped transmissions

Tilton Engineering
P.O. Box 1787
Buellton, CA 93427
(805) 688-2353 Fax (805) 688-9407
www.tiltonracing.com
  Pedal assemblies, master cylinders, brake proportioning valves, hydraulic clutch bearings, flywheels, starters

Victory Circle Chassis & Parts
700 S. Mt. Vernon Ave.
Bakersfield, CA 93307
(661) 833-4600 Fax (661) 833-4606
www.victorycirlce.com
    Race car chassis, suspension components

Wilwood Racing Products
4700 Calle Bolera
Camarillo, CA 93012
(805) 388-1188 Fax (805) 388-4938
www.wilwood .com
    Disc brake systems, hubs, brake pads, master cylinders,
pedal assemblies

Woodward Machine Corp.
P.O. Box 4479
Casper, WY 82604
(307) 472-0550
www.woodwardsteering .com
    Power rack and pinion steering systems